YIELD, INCOME AND LIVELIHOOD IN INDIAN FARMING

Conflict and Compliances

THE AUTHORS

Prof. (Dr.) Sankar Kr. Acharya, presently,Dean, Post Graduate Studies; former Head, Dept. of Agril Extension and Director, Extension Education, Bidhan Chandra Krishi Viswavidyalaya, Mohanpur, WB, born on 6th October, 1960, started his career as Assistant Professor at BCKV in 1988 and has been in teaching, research and extension over 34 years. Research Publication: 243 in National and International Journals. Book Publication: 109 books authored / co-authored. Expert Member, Agriculture Commission; Marketing and Extension Sub Committee, Govt. of West Bengal; Expert member WWF projects in BTR (Buxa Tiger Reserve Project); Expert member; DFID project on Primary Education (DPEP); Rapid Environment Impact Analysis, Visited Italy, France, Germany, China , Sri Lanka and Bangladesh. Delivered 31 Key note addresses, 13, invited and lead speakers in International Congress, Chaired 19 Sessions in International conferences, and Editors of a score of national and international journals. He has so far been awarded with 15 best paper awards in National and International Conferences Organizing Secretary of two International Conferences, held in the years, 2016 & 2018, Jointly Organized by Department of Agricultural Extension, BCKV, WB and Krishisanskriti, New Delhi. He has also been honoured to be selected as the Convener of Panel (PE-32) entitled The hunger, poverty and silence, of IUAES, University of Manchester, UK, 2013. He has successfully guided 19 Ph D scholars, of which one awarded with DAAD fellowship, Germany and two other placed as visiting scholars at Ohio State University and Cornel University, 2020-21, USA; 83 M Sc students from BCKV CU, KU, Vidyasagar University.

Dr. Debashis Mazumder , Born on 8th November, 1990 , is presently associated with, Humanitarian aid for poverty eradication foundation , KOLKATA. He did his B.Sc. (Ag.) from Doon (P.G) College of Agriculture Sc. And Technology, Dehradun in 2017. M.Sc. and Ph.D (Ag.) under the supervision of Prof S K Acharya from the Dept. of Agricultural Extension, Bidhan Chandra Krishi Viswavidyalaya , Mohanpur, West Bengal in 2019 and 2023 respectively. He attended two international seminars and one workshop ,published 5 research articles/ papers in reputed National and International Journals.

Mr. Monirul Haque, born on 9th December 1995, in Falakata of Jalpaiguri district, which is presently in Alipurduar district of West Bengal, obtained his B.Sc. (Agriculture) Honours from Palli Siksha Bhavana (Institute of Agriculture), Visva-Bharati, Sriniketan, West Bengal, in the year 2017. He obtained his M.Sc. (Ag.) in Agricultural Extension in the Department of Agricultural Extension from Bidhan Chandra Krishi Viswavidyalaya, Mohanpur, West Bengal, in the year 2019. He qualified the National Eligibility Test for Assistant Professor conducted by UGC and ICAR-ASRB in his very first attempt in the year 2019. Additionally, he has published several research articles in reputable journals. Currently, he is pursuing his Ph.D. in the Department of Agricultural Extension at the same university under the supervision of Prof. S.K. Acharya. He has been awarded the Maulana Azad National Fellowship for Minority Students to support his doctoral degree program.

YIELD, INCOME AND LIVELIHOOD IN INDIAN FARMING

Conflict and Compliances

by

Sankar Kr Acharya
Debashis Mazumder
Mr Monirul Haque

2025

Scholars World
A Division of
Astral International Pvt. Ltd.
New Delhi – 110 002

ISBN: 978-93-5919-726-5

Published by : **Scholars World**
A Division of
Astral International Pvt. Ltd.
– ISO 9001:2015 Certified Company
4736/23, Ansari Road, Darya Ganj
New Delhi-110 002
Ph. 011-43549197, 23278134
E-mail: info@astralint.com
Website: www.astralint.com

VIDYASAGAR UNIVERSITY

Professor Susanta Kumar Chakraborty

Vice-Chancellor

PREFACE

We are passing through a very difficult time and the discernible challenges are increasing productivity in agriculture with an isochronous restoration of ecosystem services. By 2050, India needs to scale up her agricultural production from a present level of 320 million tons to 550 million tons to feed the projected population size of 1600 millions, which is certainly humongous. Again, at the global level, it is estimated that an escalation of at least 60 per cent of food production from its present status, to feed the projected population of 9.5 billion.

Productivity, income, livelihood, and ecological resilience in agriculture are not performing in perfect orchestration. They are in apparent contradiction and disruption. To produce 1 kilogram of rice, the equivalent water foot print is about 4000 litres. It is obvious that enhancing food productivity, there could be more disruption and stress on soil, water and bio-diversity. The ecological resilience is a subject of musical chaos and symphony amidst ecological function-stress conservation- extraction-. Without depletion and disruption, is simply impossible to initiate and foster agriculture production as well as productivity. But the question is what could be level of endurance? Wherein restoration and retention, ecological resilience will move in resonance and complementary manner The present research based book envisages as to how and whether the following factors *viz.* productivity, income, livelihood and ecological resilience can be optimally integrated on a mutually agreed upon platform of an operating agro-ecosystem There is need to elucidate and estimate the kind, nature and impact of contributory factors on the very plank of ecological functions so that productivity will generate livelihood and income and maintain the resilience simultaneously. There should also be a policy implication as an expected outcome of the research to drive agricultural production to become ecologically viable and economically productive.

The book entitled. *Yield, Income and Livelihood in Indian Farming: Conflicts and Compliances* by Prof. Sankar Kumar Acharya and Dr. Debashis Mazumder has been a unique publication considering its relevance, both at national and international levels. Prof Acharya, a prolific writer and communicator, has chosen the much needed book authorship, in the present global context, to feed the academia, scholars and policy makers in dealing with the issues interrelated with ecology, economy and sociology.

(Professor Susanta Kumar Chakraborty)

Midnapore 721102 West Bengal India

Telefax: (03222) 275329 (Office)

E-mail: vc@mail.vidyasagar.ac.in

CONTENTS

Preface v

1. **Introduction** 1

2. **Concepts and Theory** 5

Theoretical Orientation 5
Defining Farmer, Farm and Farming System 5
Importance of Rice for Human Life 6
Calorie content 7
Current Scenario of Rice Farming in the World 7
Global Rice Production and Consumption 8
Resource Harvesting from the Pond: Hook and Line, Lula and Coastal Harvest 9
Linking Livelihoods and Poverty 10
Social Ecology of Hunger, Poverty and Livelihood in Indian Agriculture 10
Non-farm Sector as Part of the Development Strategy 11
Linking Food Systems to the Jobs Agenda 11
Skill Enhancement and Investment in Human Capital 12
Quality of Non-farm Sector 13
The Binary of Farm and off Farm Economy 13
Sustainable Agriculture and Rural Livelihoods 15
Current Levels of Poverty and Deprivation 15
Basic Issues in Sustainability 16
Issues of sustainability in Agriculture in India 17

3. **Literature Citation** 25

Here are Some Reviews Related to Rice Farming 25
The Role of Agriculture in Development: 31
Mitigating Food Price Volatility and Ensuring Food Security 32

Farm Size and Land Tenure 33
Income Diversification in Rural Areas 33

4. Research Setting **37**
Area of Study 38
Profile of the State West Bengal 38
Profile of Nadia District 41
Profile of Haringhata Block 46
Profile of Village 51

5. Research Methodology **59**
Locale of Research 59
Sampling design 60
Pilot Study 61
Variables and their Measurement 61
Methods Used for Data Collection 65
Statistical Analysis and interpretation of Data (Analytic tools) 66
Principal Component Analysis 72

6. Research and Case Studies **75**

7. Summary and Conclusion **129**
Objectives of the Study 130
Research Setting 130
Research Methodology 130
Salient Findings 132
Canonical Covariate Analysis: Interaction between Left Side Dependent and Right Side Independent Variables 136
Conclusion 139
Glimpses of Field Interaction for Data Collection and Visualization 140

8. Future Scope of The Study **145**
Limitations 145

Bibliography **147**

Index **155**

Chapter 1

INTRODUCTION

"Farming isn't a battle against nature, but a partnership with it. It is respecting the basics of nature in action and ensuring that they continue."

Jeff Koehler

Indian agriculture now stands at a cross road. By 2050, Indian population will take over that of China's. The world population in the same year will be close to 9.5 billion. And to feed the human population of this planet, the agricultural production needs to be increased 60 per cent over the present one by 2050. For India, the target is 450 million tons by 2050. This is a humongous task. This mammoth escalation of targeted food production has to be carried out with the constraints and barrier created and imposed by climate change, erosion of soil, bio-diversity, ecological disruption due to mindless anthropogenic action and wrong planning.

The production, productivity, income, livelihood, and ecological resilience in agriculture are not performing in perfect orchestration. They are in apparent contradiction. To produce 1 kilogram of rice, the equivalent water foot print is about 4000 litres. It is obvious that enhancing food productivity, there could be more disruption and stress on soil, water and bio-diversity. The ecological resilience is a subject of musical chaos and symphony amidst ecological function-stress-conservation-extraction-. Without depletion and disruption, is simply impossible to initiate and foster agriculture production as well as productivity. But the question is what could be level of endurance? Wherein restoration and retention, ecological resilience will move in resonance and complementary manner.

The present study envisages as how to and whether the following factors *viz.* productivity, income, livelihood and ecological resilience can be optimally integrated on a mutually agreed upon platform of an operating agro-ecosystem..

There is need to elucidate and estimate the kind, nature and impact of contributory factors on the very performance of ecological functions, so that productivity will generate livelihood and income. There should also be a policy implication as an expected outcome of the research to drive agricultural production, ecologically viable and economically productive.

When this diode is being formed, the wage, income, and resilience are being maintained as an organic expansion of agricultural production, process and Management.

West Bengal is growing fast through a process spectacular transformation in it's rural life process. For the last couple of decades West Bengal kept on experiencing a change in agricultural growth at the rate of 4 per cent approx. Modernizing agriculture keeps replacing the domination of field crops with the emergence of commercial crops, fishery and livestock enterprises. All these are happening to generate more livelihood, more income and assured higher food security. West Bengal is possessing around 3 lakh ha of inland water body of which nearly 49 per cent of water body are scientifically managed or rendered commercially viable.

The homophile cropping pattern (a monotonous cropping pattern followed over a long period) keeps depleting the moisture, and some other minerals constantly and leaving a specific layer extremely depleted, ecologically fragile and fairly vulnerable to the problems like acidity, salinity, compaction etc.

The syndromes are very conspicuous in areas of highly depletive agriculture like Punjab,Haryana,Tamil Nadu,Andhra Pradesh and West Bengal as well. The indication of diminishing productivity,reducing livelihood and income redundancy are becoming prominent and alarming.

So, a diverse farming system by replacing prototype agriculture is not only be agricultural reality but also an ecological necessity to sustain all kinds of micro,meso and mega life forms.

That is why and how the study of fishery delving into the need, necessity and niche of fish enterprises imbibing ecological, economic and social milieus has been a modern approach towards redeeming the value of farming system approach to be followed specially by the manager of miniscule land resources.

In the changed global scenario and with the dent for diversification of agricultural enterprises getting competitive in the global market, management of both quantity and quality are of prime issues in fish enterprise. The present study envisages assessing the productive performance of fresh water fisheries (fingerlings) in the defined domain of yield, income and employment generation and with a specific objective of taking the fish farmers as the respondents, who are managing it primarily at a domestic scale. While productive components of fish enterprise in terms of generation of yield, income and employment has been conceived as the consequent character, the causal character being a score of agro-economic, socio-personal and management by nature.

A systems perspective provides a logical framework for managing changes in social-ecological systems. The dynamic interactions of ecological and social processes that characterize most of today's urgent problems necessitate a social-ecological framework for planning and stewardship. Any sustainable solution to a resource issue must be compatible with current social and ecological conditions and their likely future changes. A resource policy that is not ecologically, economically, and culturally sustainable is unlikely to be successful. Sustainable resource stewardship must therefore be multifaceted, recognizing the interactions among ecological, economic, and cultural variables and the important roles that past history and future events play in determining outcomes in specific situations. In addition, systems undergo cyclic changes in their sensitivity to external perturbations, so management solutions that may have been successful at one time and place may or may not work under other circumstances.

The complexity of these dynamics helps frame the types of stewardship approaches that are most likely to be successful. It is unlikely that a rigid set of rules will lead to successful stewardship because key decisions must frequently be made under conditions of novelty and uncertainty. Moreover, under current rapid rates of global environmental and social changes, the current environment for decision-making is increasingly different from pat conditions that may be familiar to managers or the future conditions that must be accommodated. The more rapidly the world changes, the less likely that rigid management approaches will be successful. By considering the system properties presented above, however, we can develop resilience-based approaches that substantially reduce the risk of undesirable social-ecological outcomes and increase the likelihood of making good use of unforeseen opportunities. This requires managing for general system properties rather than for narrowly defined production goals. In this section, we present a framework for this approach that is described in subsequent chapters.

Researchable queries are:

- If productivity changes in a positive direction, whether income and livelihood are changing in the same direction.
- In case income is changing for a farmer, whether it commensurate with productivity and livelihood.
- In this case if livelihood generation improves, whether both income and productivity are changing in a positive direction.
- Is it possible that all these three, i.e. income, productivity and livelihood are growing simultaneously and isochronously?

Objectives of the study:

- To delineate a general understanding of yield, income and livelihood of selected enterprises viz. Fish, rice and vegetables for the research locale.

- The interactions and impact of these three predicted characters viz. yield, income and livelihood in terms of a score of socio-economic and ecological predictors.
- To isolate the point of optimization or a common platform for landing wherein yield, income and livelihood will be behaving in a mutual incremental way to reach a threshold.
- To generate a micro sociological policy implication based on the empirical study.

Chapter-2

CONCEPTS AND THEORY

Theoretical Orientation

It is the scholastic and theoretical exercise that configures the operational definition and text of entire research work the conceptual inputs, logic of discourses and the pursuit of knowledge come together to frame the theory in compliance and in congruence with the set of objectives and research design spelt out in the research process.

Defining Farmer, Farm and Farming System

A farmer is a person engaged in agriculture, raising living organisms for food or raw materials. The term usually applies to people who do some combination of raising field crops, orchards, vineyards, poultry or other livestock. A farmer might own the farmed land or might work as a laborer on land owned by others, but in advanced economies, a farmer is usually a farm owner, while employees of the farm are known as farm workers, or farmhands. However, in the not so distant past, a farmer was a person who promotes or improves the growth of (a plant, crop ,etc.) by labor and attention, land or crops or raises animals (as livestock or fish). Farmers are often members of local, regional, or national farmers' unions or agricultural producers' organizations and can exert significant political influence. In the context of developing nations or other pre-industrial cultures, most context of developing nations or other pre-industrial cultures, most farmers practice a meager subsistence agriculture–a simple organic-farming system employing crop rotation, seed saving, slash and burn, or other techniques to maximize efficiency while meeting.

Farm is an area of land that is devoted primarily agricultural processes with the primary objective of producing food and other crops; it is the basic facility in food production. The name is used for specialized units, such as arable farms, vegetable farms, fruit farms, dairy, pig and poultry farms, and land used for the production of natural fibers, bio-fuel and other commodities. It includes ranches, feedlots, orchards, plantations and estates, smallholdings and hobby farms, and includes the farmhouse and agricultural buildings as well as the land. In modern times the term has been extended so as to include such industrial operations as wind farms and fish farms, both of which can operate on land or sea.

Farming system is defined as a population of individual farm systems that have broadly similar resource bases, enterprise patterns, household livelihoods and constraints and for which similar development strategies and interventions would be appropriate. Depending on the scale of analysis, a farming system can encompass a dozen or many millions of households. Here, in this study I will highlight the rice farming, fish and vegetable farming.

Importance of Rice for Human Life

Rice as a staple global food

Rice, wheat and maize are the three leading food crops in the world; together they directly supply more than 50% of all calories consumed by the entire human population. Wheat is the leader in area harvested each year with 214 million ha, followed by rice with 154 million ha and maize with 140 million ha. Human consumption accounts for 85% of total production for rice, compared with 72 % for wheat and 19 % for maize.

Rice provides 21% of global human per capita energy and 15% of per capita protein. Although rice protein ranks high in nutritional quality among cereals, protein content is modest. Rice also provides minerals, vitamins, and fiber, although all constituents except carbohydrates are reduced by milling.

The world average consumption of rice in 1999 was 58 kg, with the highest intake in some Asian countries; Myanmar has the highest annual consumption at 211 kg/person. Rice eaters and growers constitute the bulk of world's poor; according to the UNDP Human Development Report for 1997, approximately 70% of the world's 1.3 billion poor people live in Asia, where rice is the staple food.

To some extent, this reflects Asia's large population, but even in relative terms malnutrition appears to affect a substantially larger share of the population in South Asia than in Africa. For these people, rice is the most important commodity in their daily lives. In countries, such as Bangladesh, Vietnam, and Myanmar, the average citizen consumes 150-200 kg annually, which accounts for two-thirds or more of caloric intake and approximately 60% of daily protein consumption. Even in relatively wealthier countries, such as Thailand and Indonesia, rice still accounts for nearly 50% of calories and one-third or more of protein.

The effect of rice on global economy

Rice is also the most important crop to millions of small farmers who grow it on millions of hectare throughout the region, and to the many landless workers who derive income from working on these farms. In the future, it is imperative that rice production continue to grow at least as rapidly as the population, if not faster. Rice research that develops new technologies for all farmers has a key role to play in meeting this need and contributing to global efforts directed at poverty alleviation.

Agricultural population densities on Asia's rice producing lands are among the highest in the world and continue to increase at a remarkable rate. Rapid population growth puts increasing pressure on the already strained food-producing resources. The aggregate population of the less developed countries grew from 2.3 billion in 1965 to 7.8 billion in 2020.Asia accounted for 60% of the global population, about 92% of the world's rice production, and 90% of global rice consumption. Even with rice providing 35-80% of the total calories consumed in Asia and with a slowing of growth in total rice area, rice production more than kept up with demand in 2000.The largest producing countries like China, India, Indonesia, Bangladesh, Vietnam, and Thailand together account for more than three quarters of world rice production.

Apart from these there are other essential importance of rice which has a greater effect on farming and consuming community.

Nutritional Content of Rice Compared to other Staple Foods

White, long-grain rice: Raw, long-grain white rice is a relatively good source of energy, carbohydrates, calcium, iron, thiamin, pantothenic acid, folate and vitamin E, compared to maize, wheat and potatoes. It contains no vitamin C, vitamin A, beta-carotene, or lutein+zeazanthin, and is notably low in fiber.

Colored rice: Brown rice retains the bran layer (containing many vitamins and minerals as well as fiber); as this has not been polished off to produce white rice. Red rices are known to be rich in iron and zinc, while black and purple rices are especially high in protein, fat and crude fiber. Red, black and purple rice get their color from anthocyanin pigments, which are known to have free-radical-scavenging and antioxidant capacities, as well as other health benefits.

Calorie content

The calorie content of 1 cup of cooked rice varies from a high of 241.8 kcals for medium or short-grain white rice, to 218.4 kcals for medium-grain brown rice, 216.5 kcals for long-grain brown rice, to a low of 165.6 kcals for 'wild rice'.

Current Scenario of Rice Farming in the World

Global rice production more than tripled between 1961 and 2010, with a compound growth rate of 2.24% per year (2.21% in rice-producing Asia).Most of the increase in rice production was due to higher yields, which increased at an annual average rate of 1.74%, compared with an annual average growth rate of 0.49% for area harvested. In absolute terms, paddy yields increased at an annual average rate of 51.1 kg/ha per year, although this rate of increase has declined in both percentage and absolute terms. Rice is grown in more than a hundred countries, with a total harvested area of approximately 158 million hectares. Producing more than 700 million tons annually (470 million tons of milled rice).Nearly 640 million tons of rice are grown in Asia, representing 90% of global production. Sub-Saharan Africa produces about 19 million tons and Latin America some 25 million tons. In Asia and sub-Saharan Africa, almost all rice is grown on small farms of 0.5" ha.

Yields range from less than 1 t/ha under very poor rain fed conditions to more than 10 t/ha in intensive temperate irrigated systems. Small, and in many areas shrinking, farm sizes account for the low incomes of rice farm families. Rice grows in a wide range of environments and is productive in many situations where other crops would fail.

The highest rice yields have traditionally been obtained from plantings in high-latitude areas that have long day length and where intensive farming techniques are practiced, or in low-latitude desert areas that have very high solar energy. South-western Australia, Hokkaido in Japan, Spain, Italy, northern California and the Nile Delta provide the best examples.

Global Rice Production and Consumption

The world's largest rice producers by far are China and India. Although its area harvested is lower than India's, China's rice production is greater due to higher yields because nearly all of China's rice area are irrigated, whereas less than half of India's rice area is irrigated. After China and India, the next largest rice producers are Indonesia, Bangladesh, Vietnam, Myanmar and Thailand. These seven countries all had average production in 2006-08 of more than 30 million tons of paddy and together account for more than 80% of world production. Despite Asia's dominance in rice production and consumption, rice is also very important in other parts of the world. In Africa, for example, rice has been the main staple food-defined as the food, among the three main crops, that supplies the largest amount of calories-for at least 50 years in parts of western Africa and for some countries in the Indian Ocean. In these countries, the share of calories from rice has generally not increased substantially over time. In other African countries, however, rice has displaced other staple foods because of the availability of affordable imports from Asia and rice's easier preparation which is especially important in urban areas. On balance, in Africa, production has grown rapidly, but rice consumption has grown even faster, with the balance being met by increasing quantities of imports. Western Africa is the main producing sub-region, accounting for more than 40% of African production in 2006-08.In terms of individual countries, the leading producers of paddy(2006-08) are Egypt (7.0 million t),Nigeria (3.8 million t), and Madagascar (3.2 million t).

In Latin America and the Caribbean, rice was a preferred pioneer crop in the first half of 10th century in the savannas of Brazil, Bolivia, Uruguay, and Venezuela, and in forest margins throughout the region. Today, rice is the most important source of calories in many Latin American countries, including Ecuador and Peru, Costa Rica and Panama, Guyana and Suriname, and the Caribbean nations of Cuba,

Dominican Republic, and Haiti. It is less dominant in consumption than in Asia, however, because of the importance of wheat, maize, and beans in regional diets. Brazil is by far the largest producer, and it accounts for nearly half (46% in 2006-08) of paddy production in the region. After Brazil (11.6 million t), the largest producers are Peru and Colombia (2.5 million t each in 2006-08), followed by Ecuador (1.63 million t).

Resource Harvesting from the Pond: Hook and Line, Lula and Coastal Harvest

Women play an important role in fisheries not only as supporting the fishery through other forms of productive and reproductive labor, but also as active participants in fishing related activities. Women's work, as it relates to fishing is not explicit within the self-identified primary tasks but women do play a significant role in resource harvesting and fishing related activities. Throughout the study it became apparent that while women do not dominate work on the cerco (although a small number do participate) or as embarcados, women do play an important role in hook and line fishing (done mainly for household consumption) and lula jigging. Women also participate in large numbers in net making and in smaller numbers in coastal harvesting. As mentioned in Chapter 3, fishing activities have been divided into five categories. These categories were established in the field and include all of the self -declared fishing activities within the community whether or not they were declared as 'primary' and include: cercos, embarcados, gill nets, lula jigging, and hook and line fishing.

As noted, men dominate the commercial and professionalized forms of fishing (cercos, embarcados, and gill nets), while women play a particularly important role in lula jigging and subsistence fishing using hook and line (see Figure 4.5). Children are also important participants in the female dominated categories in fishing. Many key community members including fishers, community organizers and tourism operators describe lula jigging as financially lucrative, fun and intrinsically important. They expressed the importance of lula in both economical terms and in terms of the food security it provide to both community residents and tourism operators. It is an important time of year and in the month leading up to the start of the season it was discussed often.

Findings related to income from wage/salaried employment

- 62% and 58% of households earning from wage/salaried employment participate in Agricultural and fishing industry in Kharif and Rabi respectively.
- More households participate in agricultural sector in Kharif than in Rabi while more households participate in Construction in Rabi than in Kharif. Construction is the most important source of wage income for farm households after agriculture.

Linking Livelihoods and Poverty

A huge body of literature has demonstrated the links between households' poverty levels and their livelihood strategies in a number of different contexts (e.g. Gautam and Andersen, 2016, Martin and Lorenzen, 2016, Oumer and de Neergaard, 2011, Van den Berg, 2010, Cramb et al., 2004). For example, studying two rural districts in Uganda, Smith et al. (2001) found that 'very poor' people, who lacked the means of engaging in any economic activity other than begging or wage laboring, were unable to diversify, while the highest wealth group chose to concentrate on one or two main activities, mostly agriculture related. In contrast, the 'poor' and 'average' income groups engaged in a wide range of activities, such 27 as small-scale service enterprises, fish trading, livestock production and sale and farm laboring, to supplement their subsistence crop production. Conversely, poorer households found it difficult to diversify away from subsistence agriculture due to their relative lack of capital and access to credit. They also had few opportunities for wage labor and off-farm self-employment. The authors emphasized that liquidity or cash in hand is central to determining diversification and income growth and hence, argue that contrary to popular belief, household poverty does not lead to greater diversification

Social Ecology of Hunger, Poverty and Livelihood in Indian Agriculture

Narratives around poverty, hunger, food security and nutrition—largely subsumed in the food system—are intrinsically linked to the development of the rural economy. Rural economic structure is constantly reshaped by forces of urbanization, expanding markets, returns to livelihood opportunities, changes in land use patterns and the inherent socio-demographic structure of villages. Also, the reliance on cultivation as the main source of livelihood in rural areas is declining with the growth of smaller towns and non-farm livelihood opportunities. Thus, the future of agricultural work will look very different from what we have seen

The theory of structural transformation suggests a decline in agriculture's share in total output and employment over time. As people move out of agriculture, household income and access to non-farm economic opportunities—rather than just farm-level production diversity or farm incomes—become equally important predictors of household food and nutrition security. This is driven by the fact that, as agricultural systems modernize and markets develop, there is an increasing separation between the production and consumption decisions of As a result, home consumption declines and food security concerns progressively become an issue of access rather than availability.

To enhance access to nutrition and food, when most households progressively become net consumers of food, household income becomes the 48 most important instruments for improving welfare. The logical question therefore ensues: how can one increase income earning opportunities in rural areas such that it increases food security and welfare? The development world has taken note of the fact that income diversification is the key to rural development, poverty reduction and food security and the same applies to India as well. Along the path of structural transformation, the non-farm sector in India has gains in prominence becoming an important pathway for increasing food security (Pingali & Sunder, 2017).

In the last four decades, Indian rural output has increased by almost seven times—`3,199 billion to ` 21,107 billion at 2004–05 prices—but the share of agriculture in rural income has reduced from 72.4% to 39.2%. Pathways from agriculture to nutrition mostly assume farming—as a source of income and food—to be the most important means to access food in rural economies. However, a greater share of households in rural India now rely on markets to access food.2 Nationally representative data suggests that 88% of farming households rely on some form of non-farm income sources to sustain their livelihoods. Non-farm income is therefore becoming an important source of food security and dietary diversity Livelihood and income diversification out of farming have been considered as desirable for enabling greater structural transformation. By moving underemployed agricultural labor towards non-farm economic opportunities, it is expected to enhance capabilities and raise household living standards. During times of distress such as weather shocks, non-agricultural labor supply is generally found to increase enabling income diversification and consumption smoothening (Ito & Kurosaki, 2009).

Non-farm Sector as Part of the Development Strategy

Livelihood diversification and non-farm employment are important levers for rural economic growth. Across developing countries, the success of the Green Revolution led to the idea of a "unimodal" agrarian structure. It was believed that agricultural growth through productivity-enhancing strategies could generate economy-wide growth multipliers, leading to across-the-board income growth and employment generation. While agricultural growth did propel growth and structural transformation in many countries, demographic pressure, preponderance of small farms, declining share of household income from agriculture and commercialization have changed the role of agriculture in future economic growth. It recognizes a growing differentiation within the agricultural sectors of developing countries. Imagining the future of agriculture, based upon experience in the last 100 years, the development strategies for rural areas should now prioritize poverty reduction as part of long-term agricultural growth strategy and foster an environment of gainful job creation. Agricultural issues should be focused towards smallholders, specifically increasing their commercially viability through connecting them to market, Hazell (2018)

Linking Food Systems to the Jobs Agenda

Food systems extend far beyond agricultural land and production and provide food, energy and nutrition to the population. At the same time, they also serve an economic and social role through enhancing household access to food. Access comes through income and better jobs. As the food system cut across agriculture, health and nutrition, poverty and the environment, it can also be leveraged to create jobs. Most new jobs in the rural areas would be created in the non-farm sector. As the agricultural value chains develop, there would be greater demand for those who can work in related logistics, from aggregation to storage and processing. With the right set of skills to youth, these sectors could become a major source of job creation. Similarly, rural employment which links to agricultural inputs and mechanization are expected to develop fast with the spread of technology and cell phones. Hello Tractor in Nigeria is a stellar example of leveraging technology to generate employment avenues in rural areas. Small farmers who cannot afford tractors use this Uber-like facility (on line booking of services) for on-demand temporary access to tractors. An aggregator in the village arranges for these demands. In India, newer organizations like Gold Farm are using Farming as a Service (FaaS) model where farm equipment can be hired (custom hiring) through cell phones or call centers.

Emerging modern food value chains, which include storage, processing, distribution; transportation and retail at the mid-stream and food preparation, and restaurants at the end-stream offer multiple avenues to create gainful employment. These are also the sectors where youth and women can be employed in larger numbers. In a recent report, Future of Food: Shaping the Food System to Deliver Jobs, by the World Bank, some of these aspects are highlighted in the developing world context. This is especially true for villages in the vicinity of cities or towns. Planners should think of creating agriculture hubs (e.g., processors, agro-industries, storage, packers) for structured food value chains. This would enable an easier transportation of food to the urban population while generating downstream employment concerning moving agricultural produce off the farm and into or through the supply chain. With rural transformation, the size of agribusiness and agricultural value chains increase relative to the farm economy in relatively more developed countries of Asia and Latin America, it is about two to three times the size of agriculture; across the developed world, it could be more than ten times as large. In the post-harvest season, when the labor demand in agriculture is low, agribusiness and food value chains offer significant avenues for employment growth. It has been shown that 10% growth in organized food processing leads to 5% employment growth in this sector (World Bank, 2015). As per capita incomes increase and eating patterns shift, the demand for jobs in these off-farm segments of the food system will increase. Capitalizing on the same could employ the skilled as well as semi-skilled youth in the hinterlands.

Skill Enhancement and Investment in Human Capital

A challenge for policy makers in India, however, has been to provide skills and increase the workforce participation of rural women. A stylized fact is that most rural women in India work on their household farm as unpaid labor. According to the latest census figures, workforce participation rate of rural women is only 30% compared to 53% for rural males. Another fundamental feature of women in rural India is their low human capital. On 58% of the rural women are literate compared to 77% of rural men. This implies that almost one-fourth of rural non-farm workers in India are illiterate. Lack of education and the required skills inhibit a smooth transition into the non-farm sector. Since construction sector does not require many skills, this sector provides the greatest share of non-farm work. The formal service sector which provides a better quality of employment and written contracts, however, requires more skilled and educated workers. Skilled and educated workers, however, are short in supply. This is a major problem with generating non-farm employment which stems from poor schooling quality in early childhood to lack of vocational training post formal schooling years. While India prides itself in achieving a near-universal school enrolment ratio, it has an abysmal record of actual learning outcome of children (Kingdon, 2007). For the firms, the shortage of skilled labor stems in two ways: lack of a sufficient number of trained personnel and trained people lacking in the required job skills (Mehrotra, 2014). This has been attributed to the outdated syllabus at these vocational institutes which create a mismatch between the quality of training and requirements of the job.

Quality of Non-farm Sector

The quality of jobs being created in the non-farm sector can be dubbed as ordinary—informal and casual—regarding their potential for rural transformation and lowering structural poverty. The poor quality of rural nonfarm sector jobs especially for women leads to greater withdrawal from the labor market Poverty, vulnerability to poverty, and informal employment status are highly associated. Around 79% of workers who work in the informal sector can be classified as poor without any job or social security. These workers not only work at low wages, but their working conditions are also miserable. Informality is not particularly specific to rural areas. Even for urban dwellers and migrant, employment and jobs opportunities for the low-skilled workers is nothing but precarious (Breman, 2016).

There have been clear complementarities between the development of agricultural and non-agricultural sectors (Foster & Rosenzweig, 2007). As the agricultural share in the GDP declines, rural non-farm economy becomes a conduit for the resource flows from agriculture to other sectors. Nonfarm economic activities are therefore central to the overall processes of economic growth and changes in the food systems.

The Binary of Farm and off Farm Economy

Changing economic structure, livelihood opportunities, urban growth opportunities and the commercialization of agriculture—in the course of the structural transformation process—lead to the greater role of markets in accessing food. As a result, means to household food access in the future would be influenced by earning capacities. Farming households would rely on the market value of their produce to access food, whereas the earning capacity of the households dependent upon the non-agricultural sector for rural livelihood challenges: moving out of agriculture. The blurring of the rural-urban distinction and greater ease of commute are expected to diversify the portfolio of economic opportunities available to the rural households. Greater fragmentation of landholdings would further increase the role of nonfarm sector in facilitating labor movement out of agriculture. This process would lead to a change in the profile of agricultural workers, and economic activities would increasingly become wage-oriented similar to the experience of the developed world. Rising urbanization and urban consumption demand would create fertile opportunities for the non-agricultural workforce to migrate, commute and get employment in food value chains.

Sustainable Development Goals (SDGs)—zero hunger and a poverty-free world: It depends crucially on how rural areas transform and the nature of their inclusiveness. Given the slow pace of rural transformation in India, the potential for leveraging food systems to propel the growth of the nonfarm sector is immense. Through organized upstream and downstream networks of activities with the emergence of greater agribusiness opportunities, surplus agricultural labor—especially youth and women—are expected to be employed in value chain processes. Such inclusive transformation of rural spaces—by including those who are left out of non-farm employment—is essential to reduce rural poverty. The other important channel for propelling agricultural growth is to strengthen the rural-urban continuum which provides ample opportunities to the small farmers and other rural population to share in the fruits of urban economic growth. Till now, policy makers in India have not recognized the potential of small towns and the peri-urban spaces to create job opportunities. Recognizing these newer urban settlements and providing them with urban amenities could be a catalyst for non-farm diversification

Local agglomeration economy benefits could be realized through creating clusters of urban-rural spaces which feed the consumer services demand for agricultural households as well as the market for new inputs, technologies and information. The rural-urban continuum would create alternatives to less remunerative migration often resorted as a strategy to escape poverty and hunger and benefit from agglomeration effects from the poorer regions. While urbanization and changing employment patterns offer opportunities for a more diversified food system, the challenge lies in ensuring these transformations are smooth and contribute to sustainable poverty reduction. Most importantly, the quality of human capital is key to change. It has been well documented that India's economic growth has not been able to generate sufficient employment in the manufacturing sector. Similarly, quality service sector jobs require specific skills depending on emerging job requirements; labor transition into the non-farm sector is difficult for most of the farm-based labor. Despite achieving universal enrolment in primary schools, learning outcomes remain low. Similarly, the lack of vocational educational facilities further limits the opportunities to transition into more gainful and formal employment avenues. These are some of the structural issues which have held back the transformation of India's economy in general.

Sustainable Agriculture and Rural Livelihoods

For the developing countries like India, where a majority of families, in both the farm and non-farm sectors, derive their livelihoods from agriculture, sustainability of agriculture cannot be discussed or even defined in isolation of the issue of livelihoods. Livelihood is defined as adequate stock and flow of food and cash with an individual or a family to meet its basic needs. Livelihood security then means secured ownership of, or access to, resources and income-earning activities, including reserves and assets to offset risks, ease shocks and meet contingencies. There are four principal ways of acquiring livelihoods by the rural households. First is the production-based livelihood. A large proportion of the small and marginal farmers gain livelihoods through production on small pieces of land. For these households, availability or access to inputs and improved methods of production are quite critical for their livelihoods. Second is the labor-based livelihood. Most of the small landholders and landless rural households derive livelihoods by selling their labor. For their livelihoods, demand for labor, wage rates and prices of food are the critical factors. Third is the exchange- or market based livelihood. Those rural households which produce surplus food and non-food agricultural products or non-farm goods earn their livelihoods by selling these surpluses in the market. The marketing system for these products and relative prices of what they sell and what they buy, affect their livelihoods. The fourth set of livelihoods is transfer-based entitlements. The households without any income-earning asset or able-bodied person to work depend for their livelihoods on transfers from the government or other social organizations. Government's social security and food assistance programmes are relevant for this group of rural households in fulfilling their livelihood requirements. The rural livelihood systems in the developing countries thus encompass a broad range of factors and depend on several macro subsystems of the economy.

Current Levels of Poverty and Deprivation

While discussing sustainability of agriculture, the fact that poverty continues to be a major problem in many parts of the world cannot be ignored. If poverty line is defined as one dollar per day, nearly 1.2 billion people live below this income. Poverty is more pronounced in the rural areas. Seventy-five per cent of the total poor live and work in the rural areas. Nearly 90 per cent of the total poor live in Asia and Sub-Saharan Africa (SSA). Asia accounts for about two-thirds of the total world poor, with South Asia accounting for 43 per cent. A significant proportion of the rural poor live in less favored areas that are challenged by difficult agro-climatic conditions such as poor soils, low and unstable rainfall, steep slopes, short growing season and inadequate infrastructure and support services. Most of the poor people are also food-insecure. (IFPRI, 2001)

As regards India, though there is a debate on the incidence of poverty, it is absolutely clear that (a) the poverty ratio has been declining over time; (b) the absolute number of poor continues to be high; and (c) malnutrition is very high among not only children but adults also. The number of poor, which hovered around 321 million during the 1970s and 1980s, declined to 260 million during 1999-The poverty ratio, defined as the number of poor as a percentage of total population, however, declined continuously from 54.9 per cent in 1973-74 to 44.5 per cent in 1983, 36.0 per cent in 1993-94 and further to 26.1 per cent in 1999-00. Spatial distribution of poor in India shows that 48.1 per cent of them live in three states, viz. Bihar, Uttar Pradesh and Madhya Pradesh. Maharashtra, Orissa and West Bengal account for 23.5 per cent of India's poor. These six states together account for 71.6 per cent of India's poor. The studies show that the poverty ratio is higher among households in the low-rainfall and un-irrigated areas than others. Across rural occupational categories, the incidence of poverty is the highest among agricultural labor households, followed by other labor households, self-employed in non-agricultural activities and self-employed in farming. Within each of these occupational groups, incidence of poverty is higher among owners of tiny landholdings as compared to landless households, which raises a fundamental question whether ownership of tiny holdings is a boon or bane.

Basic Issues in Sustainability

As already mentioned, sustainable development basically implies attainment of the objectives of inter-generational equity. Given the extent of poverty and food insecurity, it can be argued that there is some tradeoff between alleviating poverty from the present generation and caring for the future generations. The main issue is, therefore, setting up of priorities between the welfare of presently deprived families and sustainability of resource-use. The problem of priority setting gets further compounded due to the existence of very large (and further growing) inter-personal, interregional and inter-country inequalities in the livelihoods of the present generation. The questions that have increasingly become relevant are like: Who should set the priorities? Needs of which sections of the present generations should be accorded priority over other sections and over future generations? Who should define and how to define the needs of present generation that should be forgone in the interest of the sustainability? If entitling of adequate food and nutrition to say 30 per cent of the population involves some degree of resource depletion, should these be deprived of adequate food in the interest of future generations (and of whom)? There is also an important question of projections of the needs of future generations, their lifestyles and future technologies. Can such projections be done with reasonable degree of precision? Such questions clearly reflect that political economy will not allow the neglect of present demands of the poor in pursuance of better future for others (Vyas and Reddy, 1996)

The incidence of poverty is very high among tribal societies. Out of all the societies, tribal lives closest to the nature and environment. Their access to markets is limited. Development initiatives for these societies and regions have received considerable attention and are being debated upon. In this connection, the late Pandit Jawaharlal Nehru once said, "Nothing should be denied to the tribal on the ground of preserving their culture". Dalai Lama has also said, "No matter how attractive a traditional rural society may seem its people should not be denied the opportunity to enjoy the benefits of modern development". As regards the issue of poverty reduction vs. environmental preservation, Smt. Indira Gandhi, addressing the World Forum, had said, "Poverty is the worst polluter of environment", which implies that a development course which alleviates poverty is not the cause but cure of environmental problems.

Issues of sustainability in Agriculture in India

Food and Agriculture Organization (FAO) has defined sustainable agriculture as the management and conservation of resource base and the orientation of technological and institutional changes in such a manner that ensures attainment and continued satisfaction of human needs of present and future generations. It follows that sustainable agriculture is that path of agricultural development, which is environmentally non-degrading, technologically appropriate, economically viable and socially acceptable' (FAO, 1991).

It must be recognized that agriculture by definition is the most aggressively managed ecosystem, which is closely linked to the world's food system. If the alternative to agricultural sustainability is the collapse of the world's food system, there is definitely no compromise. However, in predominantly rural economies like India, growth of agriculture is critical to the achievements of goals of poverty reduction and household food-security. This requires resolution of the issue of trade-off between sustainable agriculture and a growing agriculture. What is needed is a sustained growth of agriculture. While sustainable cropping and farming systems, recent trends in profitability of farming and some new approaches like organic farming will be the focus of discussion in the Conference, there are some other areas which need increased attention for achieving sustained growth of agriculture coupled with improved livelihood systems in the country. The strategic approach to sustained growth of agriculture that helps in improving rural livelihoods should encompass the following:

- The developing countries, like India, cannot and should not ignore the fact that the priority goal of agricultural development ought to be the removal of hunger, lifting all above the poverty line, and removing malnutrition among children.
- There is an on-going debate between protagonists of agricultural development and environmentalists. While those who give precedence to the removal of hunger and food insecurity suggest that irrigation facilities should be expanded further, environmentalists argue that water use in agriculture should be brought down. However, the best course to reconcile the conflicting objectives is to adopt a strategy which increases water-use efficiency, measured as maximum biomass per drop of water. This would require technological changes in terms of low-water using varieties of crops and adoption of such nutrient management and agronomic practices,

which reduce water-use. Sometimes, it is argued that the main problem with current agricultural development paradigm is that it is a shift away from the traditional system of water and natural resource management. In this context, it should not be forgotten that current socio-economic environment is distinctly different from an environment in which traditional system was evolved and had worked. These days the conflicts in the use of natural resources like water have increased manifold. The conflicts arise because the limited amount of water that has been harnessed needs to be shared between individuals, between sectors, between states, between countries and even between man and other living beings (animals and birds

- India receives about 4000 billion cubic meters (BCM) of water annually in the form of precipitation/rainfall. Nearly 75 percent of this is received within 100 hours. Harnessing water through either surface or subsurface water storage structures is, therefore, inevitable. However, so far, only 1900 BCM of water is utilized for irrigation, which accounts for around 90 per cent of total water utilized in the country. Nearly one-third of the total water is received in the Ganga, Brahamputra and Meghna basins. Rest of the country's area receives two-thirds of total precipitation in India. Given such an uneven distribution of rainfall, floods in some regions and droughts in several other regions are bound to occur. There is, therefore, a basis for launching a long-term programme of linking different river basins of the country to tackle the problems of floods and droughts for achieving the twin goals of sustainable agriculture and improved rural livelihoods.
- A related issue is the management of available water resources. Within the next two decades, it is predicted that India will face absolute water scarcity, deepening 'water poverty' further. It has been generally accepted that India, and other developing countries in Asia and Africa, can respond to water scarcity and the resultant water poverty facing their people, by embracing integrated water resources management (IWRM). IWRM is considered as a package of best practices for improved management of water resources with strong emphasis on direct demand-side management. Though, conceptually, there cannot be any argument against IWRM, at the operational level, it is being argued that it is not feasible in most of the developing societies. The instruments of IWRM package include (a) national and state water policies to guide all the players; (b) a water law and regulatory framework; (c) treating water as an economic good and appropriately pricing it; (d) creation of tradable water rights; and (e) participatory management of water resources. The sustainability or feasibility of this model has been questioned on the ground of degree of formal economy. While this model can work well in rich, modern and formal segment of water economy like urban areas and canal command areas, where informal economy is preponderant, workability of IWRM

is in doubt (Shah and Koppen, 2006). Nearly 80 per cent of India's rural households self-supply their domestic water requirements (NSSO, 1999) and are not connected with any formal water provider. In the case of irrigation, while farmers served with canals can be assumed to be connected to 212 Agricultural Economics Research Review Vol. 19 July-December 2006 'formal water economy', NSSO (2002) survey shows that 80 per cent of villages use irrigation mostly from wells, tanks and streams without being connected to or with any administrative system. Making direct demand management work in this situation is considered closer to impossible. Shah and Koppen (2006) have argued that the IWRM paradigm must not be allowed to obfuscate the key priorities of India, which is making good, sensible investments in improving water infrastructure and services; and making these investments work. As the world's largest user of groundwater, India's water economy has a unique dynamics of its own that demands a unique strategic response. Only location-specific and context-specific institutional models would be sustainable.

- Consumption patterns of a large section of the population (middle and upper middle class) are changing rapidly, leading to a substantial increase in resource intensity of consumption. However, resource intensity of consumption continues to be considerably lower in India than developed countries. The per capita use of man-made energy is very low. Days of sunshine, hours of breathing in open air and per capita consumption of raw/fresh vegetables and fruits is very high in India. Nevertheless, the rising trend in resource intensity of consumption cannot be brushed aside for sustainability of natural resources. To address the problem, there is a need for adopting and inculcating a model based on 3 'Rs', i.e. Reduce, Reuse and Recycle. For example, there is considerable wastewater, which, if recycled can help in easing the demand for water in urban areas. It has been estimated that the wastewater in urban areas of India is around 200 billion liters, out of which only 10 per cent is treated and recycled.
- Another issue relevant in the context of sustainable agriculture and rural livelihoods is the energy security. Due to the change in life-styles, the energy needs are growing rapidly. Further, the oil supplies are becoming costlier and unstable. India's import dependence on petroleum products is very high. This, naturally, has made bio-fuels an area of increasing attention. Bio-energy offers an attractive alternative to fossil fuels. Bio-fuels not only help cope with rising energy prices, but also address environmental concerns about greenhouse gases, and offer new income opportunities to farmers. There is a very high degree of congruence between the driving forces for bio-fuel production and the poverty reduction targets embodied in MDGs. However, it requires a careful management. The development of poses risks and has the potential to result in difficult trade-offs. Though bio-fuel production would have clear benefits for the agricultural sector,

net impact on poverty and food insecurity is not clear. If production of bio energy crops requires diversion of land and water away from food crops, food prices may rise, which would be beneficial to farmers with net surplus, but poor consumers and deficit farmers would have to balance more expensive food with less costly energy. Since poor spend more on food, this is an unfavorable trade-off. Also, we need to assess the use of energy in production of bio-fuels for judging whether there is a net reduction in greenhouse gases. However, it is possible to assure a win-win outcome through a package of measures. First, develop biomass crops that yield much higher amount of energy per unit of land or water, focus on food crops that generate by-products, which can be used for bio-energy production; and grow bio[fuel crops in less-favored areas or waste lands. Second, choose scales and technologies that can produce bio-fuel on a smaller scale in the rural areas. It will create employment in the rural non-farm sector. And the third relates to the entire direction of development of sources of bio-fuels. Different crops and processing technologies lead to different environmental outcomes. For example, ethanol produced from sugarcane is reported to be competitive with today's oil prices and also has favorable energy and carbon balances. In contrast, biodiesel produced from oilseeds and ethanol produced from maize and sugar beets is reported to be less competitive and have less favorable energy and carbon balances. More research is needed on the use of cellulose feedstocks like grasses, woody parts of wheat straw, cornhusk and discarded rice hulls. It may require a new biotechnology process using high conversion technology. In this context, public sector's role in R&D becomes important because environmental and social costs/benefits are not priced in the market, and hence, bio-energy development cannot be left entirely to the private sector. India has enormous potential for liquid bio-fuels due to its diverse feedstock and contiguous landmass. There is a need for systematic research and development plan to prudently use the available land resources.

- Sustainable use of natural resources requires that environmental accounting is made an essential part of policy formulation, planning and development decision-making. There is not enough knowledge and consciousness of environmental costs at all the stages of decision-making and policy formulation. According to some estimates, between 1980 and 1999, if the cost of environmental damage is taken into account, India's economic growth rate comes to minus (-) 5.73 per cent per annum as against plus (+) 5.66 per cent estimated otherwise. The estimates of soil degradation during the 1980s and 1990s range from 11 214 Agricultural Economics Research Review Vol. 19 July-December 2006 to 26 per cent of gross domestic product. The cost of water-logging and salinity (11 per cent of area) has been estimated to be Rs 120 billion to ` 270 billion.
- As there is an element of trade-off between environmental protection and welfare loss of the present generation, valuation of individual and social

costs and benefits is quite critical in development decisions. For example, it has been found that paddy fields emit methane gas, which is environment polluting, but this may not be true in all the rice-growing areas of the world. It is being argued that methane gas emission from paddy fields is either very low or negligible in Asia. Similarly, excessive use of fertilizers is only a limited phenomenon. In several areas of India, the fertilizer-use is still at sub-optimum levels. If new technology and high-yielding practices help resource-poor farmers in improving their livelihoods and thereby reduce the incidence of hunger and malnutrition, there is a basis for encouraging rice production and popularizing these technologies further.

- A very significant aspect of sustainability of agriculture and rural livelihoods is our perception of rural and urban areas. The relationship between rural and urban areas has undergone a significant change in recent years, which is a good sign for reducing pressure on land and water resources. It has become difficult to define the borders between the two and one witnesses a continuum of farm lands, small and large towns, suburbs, informal urban settlements, peri-urban areas and urban centers. Peri-urban areas are growing at a faster pace than the cities. Cities, peri-urban areas and towns are growing because private investments tend to be concentrated in such areas. The urban population is growing at a rate higher than that of total population and the trend is likely to continue. During 2001, out of total migrants of 307 million, around 30 per cent (100 million people) migrated from rural to urban areas, mostly for employment. In addition, there were 30 million temporary or seasonal migrants in the country. It should be noted that these migrants do not include girls or women who move owing to their marriages. The migration rates are high among the most and the least educated. Seasonal migration is dominated by illiterates. There is clear evidence that rural livelihoods have been far more multi-locational than is often assumed. These people travel from the marginal areas to towns, cities and industrial centers where they find jobs in dynamic and expanding informal sector. While these jobs may be characterized as underpaid, dangerous and sometimes insecure, they are very attractive to those who come from the marginal areas where wages or earnings are very low. These jobs are not tied to the agricultural season and options of switching jobs are high. Chances of climbing up the social and economic ladder are also high. Over the years, older 'pushes' from villages such as droughts and seasonal agricultural activity are being replaced by 'pulls' of growing urban wages and expanding job opportunities. Circular migration has been helped by expansion of road networks and spread of communication technology. The motivation of migration is shifting from just 'coping' to 'accumulation'. There is ample evidence to show that remittances from migration are contributing to improved rural livelihoods. The pattern of diversified livelihood strategies reflects a dynamic process of economic, social and cultural transformation. But, the rural development optimism has tended to yield policies that aim to discourage migration and keep the poor in the countryside. As a result, the interests of migrant workers are routinely disregarded. It should be recognized that migration to urban or richer areas offers an important route out of rural poverty and also reduces pressure on agriculture. It is in this context that there is a need for a greater recognition of the contribution of migration to poverty reduction in national poverty alleviation strategies. The existing negative policy and institutional contexts impose unnecessary high costs and risks on migrants. Attempts to control or reverse the process of migration would in fact choke off a major livelihood opportunity that has become available to those living in the marginal areas. We should look for ways to support migration. These could be reforming pro-poor programmes based on residence criteria, identity cards for migrants, skill enhancement of migrants, migrant-friendly insurance schemes, and facilitating remittances. Recently, DFID-India is reported to have set up a Migrant Labor Networking Unit to upscale the successful migrant support programme under its Western India Rain-fed Farming Project, which is a welcome initiative that needs to be replicated.(Acharya, S S, 2006)

The future of food and agriculture trends

A number of global trends are influencing food security, poverty and the overall sustainability of food and agricultural systems. The world's population is expected to grow to almost 10 billion by 2050, boosting agricultural demand – in a scenario of modest economic growth – by some 50 percent compared to 2013. Income growth in low- and middle-income countries would hasten a dietary transition towards higher consumption of meat, fruits and vegetables, relative to that of cereals, requiring commensurate shifts in output and adding pressure on natural resources.

Economic growth and population dynamics are driving the structural change of economies. The decline in the share of agriculture in total production and employment is taking place at different speeds and poses different challenges across regions. Although agricultural investments and technological innovations are boosting productivity, growth of yields has slowed to rates that are too low for comfort. Food losses and waste claim a significant proportion of agricultural output, and reducing them would lessen the need for production increases. However, the needed acceleration in productivity growth is hampered by the degradation of natural resources, the loss of biodiversity, and the spread of trans boundary pests and diseases of plants and animals, some of which are becoming resistant to antimicrobials.

Climate change affects disproportionately food-insecure regions, jeopardizing crop and livestock production, fish stocks and fisheries. Satisfying increased demands on agriculture with existing farming practices is likely to lead to more intense competition for natural resources, increased greenhouse gas emissions, and further deforestation and land degradation.

Hunger and extreme poverty have been reduced globally since the 1990s; yet, around 700 million people, most of them living in rural areas, are still extremely poor today. In addition, despite undeniable progress in reducing rates of undernourishment and improving levels of nutrition and health, almost 800 million people are chronically hungry and 2 billion suffer micronutrient deficiencies. Under a 'business-as-usual' scenario, without additional efforts to promote pro-poor development, some 653 million people would still be undernourished in 2030. Even where poverty has been reduced, pervasive inequalities remain, hindering poverty eradication.

Critical parts of food systems are becoming more capital-intensive, vertically integrated and concentrated in fewer hands. This is happening from input provisioning to food distribution. Small-scale producers and landless households are the first to lose out and increasingly seek employment opportunities outside of agriculture. This is driving increased migratory flows, especially of male members of rural households, which is leading, in turn, to the 'feminization' of farming in many parts of the world.

Conflicts, crises and natural disasters are increasing in number and intensity. They reduce food availability, disrupt access to food and health care, and undermine social protection systems, pushing many affected people back into poverty and hunger, fuelling distress migration and increasing the need for humanitarian aid. Violent conflict also frequently characterizes protracted crises. On average, the proportion of undernourished people living in low-income countries with a protracted crisis is between 2.5 and 3 times higher than in other low-income countries.

Challenges

These trends pose a series of challenges to food and agriculture. High-input, resource-intensive farming systems, which have caused massive deforestation, water scarcities, soil depletion and high levels of greenhouse gas emissions, cannot deliver sustainable food and agricultural production. Needed are innovative systems that protect and enhance the natural resource base, while increasing productivity. Needed is a transformative process towards 'holistic' approaches, such as agro-ecology, agro-forestry, climate-smart agriculture and conservation agriculture, which also build upon indigenous and traditional knowledge. Technological improvements, along with drastic cuts in economy-wide and agricultural fossil fuel use, would help address climate change and the intensification of natural hazards, which affect all ecosystems and every aspect of human life. Greater international collaboration is needed to prevent emerging trans boundary agriculture and food system threats, such as pests and diseases.

Eradicating extreme poverty, and ensuring that vulnerable people who escape poverty do not fall back into it, requires action to reduce inequalities. That means addressing inequalities both between and within countries, in levels of income, in opportunities and in ownership of assets, including land. Pro-poor growth strategies, which ensure that the weakest participate in the benefits of market integration and investment in agriculture, would improve their income and investment opportunities in rural areas and address the root causes of migration.

But pro-poor growth must go beyond agriculture, by involving both rural and urban areas and supporting job creation and income diversification. Social protection combined with pro-poor growth will help meet the challenge of ending hunger and addressing the triple burden of malnutrition through healthier diets. Permanently eliminating hunger, malnutrition and extreme poverty also requires building resilience to protracted crises, disasters and conflicts, and preventing conflicts by promoting inclusive and equitable global development.

A rethinking of food systems and governance is essential for meeting current and future challenges. Vertically coordinated, more organized food systems offer standardized food for urban areas and formal employment opportunities. But they need to be accompanied by responsible investments and concern for smallholder livelihoods, the environmental footprint of lengthening food supply chains, and impacts on biodiversity. These concerns need to be addressed by making food systems more efficient, inclusive and resilient.

On the path to sustainable development, all countries are interdependent. One of the greatest challenges is achieving coherent, effective national and international governance, with clear development objectives and commitment to achieving them. The 2030 Agenda for Sustainable Development embodies such a vision – one that goes beyond the divide of 'developed' and 'developing' countries. Sustainable development is a universal challenge and the collective responsibility for all countries, requiring fundamental changes in the way all societies produce and consume (FAO, 2017).

Chapter-3

LITERATURE CITATION

Review of Literature stands for citations and interpretation of relevant researches as well as studies. This chapter delineates, describe and deduce empirical references and support to the present study based on a set of objectives. So a review of literature directs and drives inquisitive minds to go deeper to elucidate contrast or similarities of research findings to place the present research on the strong foundation of conceptual structure.

Several studies at various regional and state levels on different aspects of the rice,vegetables and fish have been conducted by researchers,academicians and research institutions. Therefore,it is quite relevant to review the available literature that has relevance in identifying the gaps that exist in the field of present research.

Here are Some Reviews Related to Rice Farming

Shukla(1996) has categorized into Cost A1,Cost A2,Cost B and Cost C.Cost A1 includes the cost of seeds, manures and fertilizers, plant protection, livestock expenses, hired human labor, irrigation charges, land revenue, interest on working capital, depreciation of fixed assets and miscellaneous expenses. Cost A2 covers Cost A1 plus rent paid for leased land. Cost B includes Cost A2 plus rental value of owned land plus interest on fixed capital minus land revenue on owned land. Cost C includes Cost B plus imputed value of family labor.

Kumar *et al.* (2005) in their study on "Technical Efficiency of Rice Farms under Irrigated Conditions of North West Himalayan (NWH) Region-A Non-Parametric Approach" stated that hill agriculture is practiced under tough conditions because of its unique character. The hill and mountain ecosystem are unique because of topographical features and climatic variations along the gradient. In general,hills receive 750 to 1250 mm precipitation; however, only about 10 per cent of the area is under irrigation in Uttaranchal hills that too confined to the lower valleys. Sub-optimal hydro-thermal regimes and shallow soil depths thwart further extension of cultivated land. Small and scattered land holdings and limited land use is also the main feature of hill agriculture. Therefore, the food produced is not sufficient to sustain for the whole year.

Chatterjee*et al.*(2013) his study examined that Rice is a principal food crop which occupies nearly a one-fourth of the gross irrigated area in India. However,the present study attempts to judge the essence of the Green Revolution in rice cultivation and it's actual reflection regarding the factor contribution over four decades in West Bengal,India. The study measures the extent of technological change in rice cultivation. The spatial change in the TFP as well as a comparative study on productivity,input use, break-up of cost components and economic return in the paddy cultivation over different size classes across all agro-climatic zones of West Bengal has been made in order to identify the most promising zone regarding technological advancement in rice cultivation.

Chowdhury & Muvva (2018) conducted a study in Godavari District of Andhra Pradesh (i) to estimate the costs and returns involved in cultivation of rice in West Godavari District of Andhra Pradesh (ii) to compare and examine the relationship among minimum support price, wholesale price, farm harvest price and farmer expected prices (iii) to study the production and marketing constraints faced by farmers in rice cultivation (iv) to make suggestions in order to enhance the effectiveness of MSP and procurement of rice by the government. Multi-stage stratified random sampling technique was employed for selection of final respondents.A sample of 120 farmers was selected from eight villages and four mandals of West Godavari District of Andhra Pradesh.Tabular analysis was carried out to study the costs and returns of rice cultivation in the study area.The cost concepts and income measures were also analyzed through conventional analysis using various cost concepts and farm income measures.The prices like minimum support price (MSP), wholesale price (WP), farm harvest price (FHP) and farmers expected price were analyzed by compound growth rates.The comparison and the relation between these prices were examined by Neronian price expectation model. The problems during production and marketing of rice were identified and analyzed using Garrett ranking technique.Finally,the suggestions in order to enhance the effectiveness of MSP and the procurement of rice by government were documented. The average cost of cultivation per hectare of rice cultivation was Rs. 74,698.21 for pooled farmers.The average per hectare total cost of cultivation (TCOC) increased with the increase in the size of the holdings from Rs. 72,200.22 on small farms, Rs. 74,356.57 on medium farmers to Rs. 77,007 on large farms indicating a direct relationship with the size of the holding.This was due to intensive use of inputs by large farmers.The proportion of variable costs accounted for a major share in the TCOC on all categories of farms.The total per hectare operational costs ranged from ` 54,342.22(73.88%) on small farms,` 55,279.57 (74.34%) on medium farms to ` 57,007 (74.02%) on large farms,with an overall average of ` 55,437.93 (74.62%)

on pooled farms.Among the different variable costs the cost of human labor was the major cost component with an amount of ` 17,950 per hectare accounting for 24.03 per cent of TCOC on pooled farmers.

GovardhanaRao,G. (2012)analyzed the price behavior of rice in Guntur District of Andhra Pradesh. It was conducted in Guntur District of Andhra Pradesh with the specific objectives : (1) to study the trends and relationships in the behavior of minimum support price,farm harvest prices and cost of cultivation of rice in Andhra Pradesh, (2) to study trends,seasonal and cyclical variations in market prices of rice in Guntur District and Andhra Pradesh and forecast the same, (3) to study the domestic and export competitiveness of rice (4) to study the farmers and millers perception remunerative price for rice.The primary data was collected from the total sample of 90 rice growing farmers and 30 rice millers to know the perception regarding the remunerative prices for rice in Guntur District. The present study relates to the agriculture year 2010-11.The secondary data regarding the MSP,FHP and cost of cultivation of rice were obtained from various issues of Agricultural Statistics at a glance,CACP, Department of Agriculture and Co-operation and Government of India.The data on monthly wholesale prices of rice in Andhra Pradesh and Guntur market were collected from various statistical bulletins published by the Directorate of Economics and Statistics at a glance of various years,DGCI &S,Kolkata,Centre for Monitoring Indian Economy (CIME) reports and FAO,database.

Vakati, Narendra (2015) stated in his study, “Economics of production and marketing of paddy in Nellore 4 District of Andhra Pradesh” that Paddy (Oryza sativa) is a member of Gramineae family. Paddy cultivation probably dates back to antiquity and had probably been the staple and the first cultivated crop in Asia. Rice is grown over an area of 163.46 million hectare with production of 718.35 million metric tonnes and productivity of 2372 kg/ha (Anonymous,2014). In Andhra Pradesh, rice is the major food crop grown in 34.40 million hectares producing 9.31 in both Kharif and Rabi seasons with an average productivity of 3.02 t/ha during 2010-11.At national level, Andhra Pradesh is contributing 12.7% of rice production with 9.08% rice area.

Barun Kumar Paul (2019) found during the field survey in Galsi-I and Galsi-II Blocks of Purba Bardhaman that many farmers do not think so. In this context,using the secondary data on the cost of paddy cultivation for the period of 1996-97 to 2015-16,this paper analyses the cost and returns from paddy cultivation in West Bengal.It is observed that the cost has increased at a very high rate.Paid out cost at current prices during the last decade has increased by 12.1 per cent per annum. However,they do not realize the right price of their produce,and consequently,the profit margin has declined.Data reveal that farmers incurred losses in paddy cultivation up to 14 times except for in 2007-08 and in 2009-10 during 2000-01 to 2015-16.they are also being deprived of getting even the minimum support prices (MSP) due to loopholes in government initiatives.Using primary data,this paper explains why many farmers think that paddy cultivation is still a profitable activity.In doing so,the research finds that they miscalculate the cost of cultivation and that is why they perceive the profit margin to be higher.

Anwar Hussain, Naeem-Ur-Rehman Khattak & Abdul Qayyum Khan (2008) conducted a study in district Swat during 2007 to make comparative cost benefit analysis of per acre rice production of different rice varieties. Primary data was collected through structured questionnaire. Five villages from three tehsils namely Kabal, Barikot and Mattawew randomly selected.A sample size of 100 farmers was used and the respondents were randomly selected. For comparison, Cost-Benefit Analysis approach was used. The total per acre rice production of these varieties was amounted to ` 40000,52500,33600,34000,30400,30400 and 68750 respectively.The same average cost amounted to Rs. 13565 was observed for all the varieties.The Benefit Cost Ratio (BCR) of variety JP-5,Basmati-385,SaraSaila,Dil Rosh-97,Swat-1 and Swat-2.

Singh (1983) stated in his study,"Different categories of farmers differ significantly in their level of education size of holding,housing pattern and farm power possession,but their existed no significant differences in age groups."

Markovic*et al.* (1990) conducted that during 3 years investigations to investigate influence on different quality of nitrogen (0 to 200 kg/ha) on the yield quality and basic nutritive elements accumulation in Cauliflower.

ICAR (1981) viewed the farming system as a combination of productive enterprises - crops, livestock, fish, poultry etc., and development of farming programme in such a manner as to utilize resources optimally for deriving maximum profits.

Byerlee*et al.* (1982) reported that, farming system research (FSR) offered promising scope to improve the traditional and subsistence farming in developing countries and defined FSR as farm research that views the farm in a holistic manner and considers the interaction in the system. The major objective of this type of study was to solve the specific problems in farming and to increase the productivity of farming systems by generating new technology which would be appropriate for farmers They stressed the need for actively including the farmers in farm researches to develop less ambitious and more practical farming systems.

Desai (1982) revealed that crops including horticulture and plantation, livestock, fisheries and forestry production are important components of IFS. In Indian agriculture food grains production is found to be the most important aspect emphasized that other components need to be included and developed as an integrated part of the farming system.

Remenyi and Coxhead (1985) described the usefulness and effectiveness of FSR in terms of multi-disciplinary and complexity of farming. He explained the five critical characteristics of FSR in terms of its relation to small holders, multi-disciplinary and holistic approach, and inclusion of farmer in research, aiming at generating viable technology and lastly testing and verifying the research results.

He also stated that, FSR is useful in the research which is conducted with the recognition and focus towards the inter dependence and interrelationship that exist among the different elements of the farming systems. The main thrust is towards the systems diagnosis, problems identification and solution formulation.

Rosario (1985) emphasized that the need for FSR which is arising because of growing disparity between experimental and farmers" yields and limited adoption of recommended technology. Due to these reasons, yields are low and increased yields do not translate into improved economic well-being. Therefore, FSR is a strategy who properly accounts for the needs of farm sector, particularly, the small farmers which constitute bulk of the farms.

Sabrani*et al.* (1985) explained the major thrust of FSR on crop livestock research. The main aims of FSR are to develop stable cropping pattern, maintenance of soil fertility, efficient utilization of land and labor and the provision of sustained cash income. According to them, future needs of FSR are to impart training, regular research, evaluation and formulation of nationwide FSR network.

Sateesh and Sharma (1988) defined farming system as the farming systems specifically refers to a crop combination or enterprise mix in which the products and/or the by-products of one enterprise serve as the input for the production of other enterprises. It takes into account the consumption needs of the family. They further opined that farming was a stochastic, dynamic, biological and open system with human or social involvement. Being primarily biological with a high degree of dependence on weather variables and changing socio-political environments, farming system was inherently more risky than other system.

Sain (1991) commented that study of farming systems is important not only from the view point of planners but from the view point of farmers also. The planners are mainly concerned with the framing of policies in the national interest while the farmers are interested in maximizing their profit. Sound resource base and unmatched development in technical break through are obvious reasons behind the efficiency of any farming system.

Sharma *et al.* (1991) viewed farming system as a set of agro-economic activities that are inter-related and interact among themselves in a particular agrarian setting. They tried to identify farming systems followed in different agro-climatic zones of Himachal Pradesh. They indicated that vegeTable:based and fruit-based farming systems are the harbinger of profitability and surplus in the agrarian economy. In each zone there is heavy interdependence of strong linkages among different systems of farming and common property resources. Hence, integrated planning for sustainable farming system development is essential.

Singh et al. (1991) conducted a study for developing most remunerative farming system in relation to the agro socio-economic conditions of western Haryana. The study revealed that mixed farming with three milching animals was always better than arable farming in all respects.

Sundaramurthy (1985) conducted research project on integrated pest management on cotton in Tamil Nadu. Wherein the per hectare net profit earned by IPM adopter was 4845 while that earned by Non-IPM adopter was 3402. Thus, additional profit of 543 per hectare earned by IPM adopter was due to the impact of IPM techniques.

Ray and Chahal (1986) observed a wide gap between the national demonstration yield and the actual yield with respect to groundnut. The gap was very high in the states of Andhra Pradesh (15.89 q/ha), Maharashtra (6.84 q/ha) and Karnataka (5.82 q/ha) poor management practices like untimely sowing, improper seed rate.

Sundaramurthy (1987) conducted an operational research project on integrated control of cotton insects in Tamil Nadu, wherein appropriate insecticides were used at action threshold level under supervision. The mean quantity of insecticides used in non-project area was found to be 2682 gm per hectare as against 2808 gm per hectare in project area. A marginally higher quantity of insecticides used in the project area was due to continued pressure of migrates of *Heliothisarmigera* from sorghum crop. The results indicated that the per hectare net profit earned by IPM farmer was 8910, while non-IPM farmer earned 7656.

Pandurangadu and Raju (1990) conducted a study on economics of pesticides used on cotton farms in Guntur districts of Andhra Pradesh. The study revealed that, an alarming rise in the cost of cultivation of cotton was largely attributed to the increased use of expensive insecticides. Hence, not only as an economy measure but also to avoid side-effects, farmers were advised to adopt integrated pest management technology.

Anonymous (1992) conducted a study on integrated pest management in Guntur and Prakasam districts of Andhra Pradesh. In both the districts, farmers used the package of plant protection recommended. They experienced an increase in yield of 3-4 quintals per acre, which represented a significant increase of more than 80 per cent. Also, the total number of application of pesticides was reduced. Educating the farmers on the right use of technology and by adoption of integrated pest management practices, Helliothis on cotton can be managed effectively.

Patil (1992) reported that in the demonstration on integrated pest management on cotton, the net profit from the IPM block was about 1029 per acre whereas that from the farmers block was minus 55 per acre, mainly due to high cost of plant protection chemicals used by the farmers in their traditional practices.

Suryawanshi and Mahindre(1993) studied the impact of viable technology for promoting oilseeds in Maharashtra. The data from frontline demonstrations laid out by the centers of All India Co-ordinated Research Project on oilseeds in Maharashtra laid out along adjacent plots of farmers following traditional practices were used to show the comparative production potentials and benefits accruing from viable technology. At the national level, the recommended technology increased the yield by 36 to 45 per cent in groundnut, 35 per cent in sesamum,

21 to 47 per cent in sunflower, 21 to 63 per cent in safflower and 77 per cent in Niger crop. In Maharashtra, the productivity on demonstration plots increased by 53 per cent in groundnut, 22 to 48 per cent in sunflower and safflower over that obtained by the farmers following traditional practices. The analysis showed that significant yield difference was due to the adoption of improved technologies. The yield difference in case of kharif groundnut, summer groundnut, sesamum, sunflower and safflower were 56, 20, 200, 226 and 640 per cent, respectively.

Muralidharan (1987) studied the resource use efficiency in rice production in Kerala employing the Cobb-Douglas production function. The adjusted R^2 was 0.84 indicating that 84 per cent of the variation in yield of paddy could be explained by the estimated production function. The coefficient of land and human labor were positive and significant at one per cent probability level.

Vishweshwar (1994) employed Cobb-Douglas production function to measure the efficiency of inputs used in the production of cotton by IPM and Non-IPM adopted farmers in Malaprabha command area in Karnataka. The study indicated that the ratio of MVP to MFC for land was greater than one, while it was less than one for labor. It was negative for seeds, fertilizers and pesticides in conventional farmers. In case of IPM adopted farmers, the MVP to MFC ratio for land, labor, seeds and fertilizers were greater than one and it was negative for fertilizers.

Sharma and Singh (1996) indicated that the feed concentrate was the most important input affecting milk production. The regression coefficients of this input were positive and statistically significant in all the equations fitted, indicating that the farmers could increase their milk output by feeding more concentrates to the animals on both the groups of households. The regression coefficients of green fodder and dry fodder were also positive and significant in most of the equations fitted.

The Role of Agriculture in Development:

De Janvry (2010). Having been a key preoccupation of developing country governments, donors and the international community during the 1960s and 1970s, agriculture disappeared from the development agenda in the 1980s and 1990s, only to reappear in the first decade of the 21st century because of neglect and underinvestment (see Figure 1). There is now renewed interest in the problems of the sector—not to a small extent, thanks to the 2008 World Development Report. For example, the G8 countries promised $22 billion for investment in agriculture during

Timmer (2002); An additional problem that arises with cross-country studies is that differences in country conditions do not allow for a general relationship between agricultural and aggregate economic growth. Factors such as openness to trade could alter the relation between agriculture and non-agriculture.

Global markets can be a substitute for what Timmer (2002) calls first-order effects of agricultural growth (because they provide international capital flows and food imports). This might explain the different magnitudes of the positive relation found by Bravo-Ortega and Lederman (2005) for Latin America compared with other regions. Hence, the importance of linkages between the agricultural sector and the rest of the economy differs across countries.

Mitigating Food Price Volatility and Ensuring Food Security

The main reason for price stabilizing efforts by governments and the use of export bansduring times of crisis is to ensure food security. Given the mixed success of price stabilizingpolicies, the question arises if such policies are necessary to prevent adverse effects of volatileprices on the poor or if other measures, such as the improvement of social safety nets, might bemore effective in helping individuals in developing countries to cope with unstable food pricesand to guarantee food security.

Although food subsidies are meant to support the poor, they are often not well targeted. And subsidies are direct interventions in the market and can lower incentives for producers (World Bank 2008). Social safety net measures such as food-for-work programs, school feeding or in-cash or in-kind transfers are more targeted instruments to protect consumers from high food prices but are less common than market interventions or trade policy measures (Demeke et al. 2008). Bangladesh, for example, has recently extended its work-for food program in response to natural disasters and higher food prices (Coady et al. 2003).

It is important that such safety net arrangements be in place before a shock hits in order to be. In that case, they can prevent households from falling into chronic poverty by helping them cope with higher food prices. However, most social safety net policies only address long term poverty and are not tailored to cope with transitory income shocks.

Hence, new, innovative approaches have to be developed for social safety nets to respond quickly to unexpected shocks (Byerlee et al. 2010).

As a positive example, recipient households of conditional cash transfers in the Progresa program in Mexico that suffered from income shocks have showed lower school dropout rates than non-recipients (Alderman 2008, de Janvry et al. 2006). This also suggests that, whereas trade policies or market interventions can only smooth aggregate shocks to food security, such as higher overall food prices, social safety net measures can also help to overcome idiosyncratic shocks to households and guarantee adequate food accessibility.

Often, however, limited government budgets do not allow for extensive safety nets. As a result, increased productivity in subsistence farming is probably the most effective way to guarantee food security for small farmers. Moreover, investments to increase productivity are long-term instruments for agriculture for development. While short-term policies to cope with transitory shocks are important, the long-term measures described above – such as productivity increases, developing more shock-resistant crops, improving farmersaccess to modern markets and creating employment opportunities in the rural non-farm sector – will be necessary to reduce the risk of shocks ex ante and guarantee food security in the future (Byerlee et al. 2010).tee food security.

Farm Size and Land Tenure

Many farmers in developing countries operate on a small scale. Increasing farm size is key for improved incomes in agriculture because it allows for the use of mechanization that has indivisibilities (with differences in access to credit by small and large farms favouring the latter), implying increasing returns to scale and higher profitability per hectare.

In a detailed empirical paper, Foster and Rosenzweig (2010) show that this is the case today in India, where most farms are too small to exploit the productivity and cost-saving advantages from mechanization. But land consolidation in larger farms with continued population growth and limited long distance migration requires vigorous employment creation in the rural non-farm economy, basically in secondary towns and cities. Agriculture-for-development strategies need to focus on the smallholder sector, understand the challenges they face and find ways to make them more productive (World Bank 2007).

New developments in the agri-business and food sector might shift the competitive advantage to large-scale producers and make it difficult for smallholders to participate actively in food markets. Two developments are of special importance: the supermarket revolution, and standards and certification. The wave of new supermarkets in developing countries has led to new procurement systems (Eastwood et al. 2010).

Income Diversification in Rural Areas

The rural non-farm sector, perceived to be unproductive and of negligible importance, had in the past received much less attention from governments in the past than, say, the issue of urban bias (Lanjouw and Lanjouw 1995, 2001). Recent contributions point out that the rural nonfarm sector can serve as a bridge between agricultural-based livelihoods and industrial ones (Barrett et al. 2010), thereby playing an important role in a country's structural transformation. As pointed out above, further agricultural growth will have to come from capital-led intensification of production, which will limit the capacity of agriculture to employ a constantly growing rural labor force. Hence, rural and peri-urban sectors will play an important part in absorbing additional labor in the future. Employment in those sectors is stimulated by agriculture growth through the production and consumption linkages described above, and is an important complement to agriculture for rural poverty reduction (de Janvry and Sadoulet 2009).

The most important roles played by the rural non-farm sector are the generation of employment and income for the rural poor. Many households in rural areas today do not specialize in either agriculture or non-farm activities but derive their income from multiple sources. Income diversification plays an important role in the livelihoods of rural populations. Farmers grow different types of crops during the year. Families derive income from non-farm rural activities and receive remittances from household members who have migrated. The literature mentions several reasons why rural households and individuals—as opposed to their urban counterpart—diversify rather than specialize. Barrett et al. (2001) name four main causes for income diversification into non-agricultural activities: seasonality in employment opportunities, diminishing returns to factors of production, market failures and risk management. Due to seasonal variations in returns to labor in farm activities, individuals might work temporarily in the non-farm sector. Some household members might work off-farm if there are diminishing returns to labor in agricultural production. Simultaneous income generation from both farm and non-farm sectors can be a strategy to cope with income risk if appropriate insurance markets are not available. Engaging in activities independent from agriculture such as manufactures can counter farm-related risk. Households in Burkina Faso that diversified their income into non-farm activities were more able to cope with droughts during the 1980s (Webb and Reardon 1992).

Food and agriculture are now back in the headlines because of poverty linked to price volatility and high food prices is on the rise. In the past few years, there have been variousattempts to address the issue of food security – including an international race for agricultural land, large private investments in agriculture attracted by favourable markets, disruptive protectionist policies in food staples, revival of large-scale subsidies to agriculture in the name of food self-sufficiency, and new commitments by donors to increase spending on agriculture after 20 years of neglect. Since the key question is how to use agriculture in support of a structural transformation of the economy, we began by examining the role played by agriculture in the development process and its interactions with other sectors. In poor countries, agricultural growth has a huge capacity to reduce poverty. Due to this potential, improving productivity in the agricultural sector in developing countries is critical and an essential step to reach the Millennium Development Goals. Some 75 percent of today's poor living in rural areas would benefit massively from higher incomes in agriculture. Moreover, agriculture also has the potential to generate economic growth in developing economies that depend to a large extent on this sector, for example in many Sub-Saharan African countries. The most difficult challenges are institutional and are related to market failures, missing markets and property rights. Agriculture can also be an engine of growth and provide employment opportunities for the rural non-farm economy because of its linkages with small cities and rural areas. Rural development and community-driven development can assist in this process. The government will need to play an important role for many of the tasks suggested in this paper. It should not, however, be the only purveyor. The private sector will be the main source of investment funds and a supplier of services. Donors, NGOs and civil society organizations (who benefit from local and external private expertise when implementing projects) will also play a key role. To identify the right mix of these actors and to establish an effective cooperation among them will be important. New approaches to increase productivity in these countries have to be found. The possibility of further land expansion to increase agricultural output will soon be exhausted and intensification will be the only way to increase productivity in the future. For its Green Revolution, Africa therefore needs high-yielding varieties that are adapted to local conditions. In addition, to guarantee adoption of such crop varieties and to integrate small farmers into modern value chains, existing barriers – such as low education, missing infrastructure, lack of credit and insurance markets – and insecure property rights have to be addressed. And new ways of information dissemination and learning methods, such as the use of communications technology in extension services, can foster adoption and profitable cultivation among farmers. Increasing productivity among smallholders in developing countries is also an instrument to guarantee food security in the long-run.

As if new evidence was required, the 2008 food crisis has demonstrated once again the vulnerability of the poor to income shocks due to food price increases. As commodity prices rise once again, it is important to adopt measures to limit volatility and put in place effective coping mechanisms. Macroeconomic approaches to stabilize prices in national markets are not promising. Policies such as social safety nets, which help the poor to cope with income shocks, have the potential to mitigate adverse effects and prevent households from falling into chronic poverty. Beggar-thy-neighbour trade policies to stabilize prices and guarantee national food security have been counterproductive: during the 2008 food price crisis, policies adopted by some developing countries have harmed poor populations and reversed some past gains. The best instruments to protect small farmers from income shocks are ex-ante measures such as increased productivity that reduce the risk of shocks in the first place (Dethier et al., 2011).

Chapter 4

RESEARCH SETTING

Research setting refers to the detail information of an area, the social-ecology, where the study has been conducted. The study area, generally comprises of a particular geographical area viz. a state, district, block, or gram panchayat area selected according to the convenience of the investigator who possess adequate knowledge regarding the location, communication facility etc. of the locality so that he/she can easily approach each and every corner of the area for data collection. Here. in this study the data collection was done mostly over telephone based on the interview schedule due to Covid-19 pandemic situation. Besides the investigators must also have the basic knowledge about the socio demographic background of the people so as to have an easy understanding of their knowledge, attitude and behavior.

A research setting is a surrounding in which inputs and elements of research are contextually imbibed, interactive and mutually contribute to the system performance. Research setting is immensely important because it is characterizing and influencing the interplays of different factors and components. Thus, a study on perception of farmer about the issues of persuasive and other topics, certainly demands a local unique with natural set up, demography, crop ecology, institutional set up and other socio-cultural milieus.

It comprises of two types of research setting viz., Macro research setting and Micro research setting. Macro research setting encompasses the state as a whole, whereas micro research setting starts off from the boundaries of the chosen districts to the block or village periphery. The notion behind this form of presentation is to internalize the study environment in terms of broader perspectives with state as reference frame and district, block profile as units for in depth study.

Area of Study

The area of investigation of this study is situated in the state of West Bengal located in the eastern part of India. The state of West Bengal in eastern India has unique social and ecological backgrounds which influence the living standard and behavioral patterns of the people in many ways. The area of the investigation comprises of five selected villages viz. Chak Birohi, Kantabele, Ghoragacha, Mitrapur and Kashtodanga of Haringhata block, Nadia district.

Profile of the State West Bengal

West Bengal (*Paschim Banga*) is a state in the eastern region of India along the Bay of Bengal. With over 91 million inhabitants, it is the and the in India. West Bengal is situated in the north eastern part of India and lies between 21°37' 27°10' north latitude and 85°51' 89°53' east longitude. Covering an area of 88,752 km^2 (34,267 sq. mi), it is also the eighth-most populous country subdivision of the world. Part of the of the Indian subcontinent, it borders Bangladesh in the east, and Nepal and Bhutan in the north. It also borders the Indian states of Odisha, Jharkhand, Bihar, Sikkim and Assam.

The state capital is Kolkata, the third-largest metropolis, and seventh largest city by population in India. West Bengal includes the Darjeeling Himalayan hill region, the Ganges delta, the Rarh region and the coastal Sundarbans. The state's main ethnic group are the Bengalis, with the Bengali Hindus forming the demographic majority.

It is flanked by the Bay of Bengal in the south, Sikkim on the north, Assam on the east and Jharkhand and Orissa on the west. It covers an area of 8.85 million ha, representing only 2.7% of the total area of the country. The State is divided into 20 administrative districts, viz. Bankura, Birbhum, Burdwan, Kolkata, Coochbehar, Darjeeling, Hooghly, Howrah, Jalpaiguri, Malda, Midnapore-East, Midnapore-West, Murshidabad, Nadia, North 24 Parganas, North Dinajpur, Purulia, South 24 Parganas, South Dinajpur and Alipurduar.

Post Indian independence, West Bengal's economy is based on agricultural production and small and medium-sized enterprises. For many decades the state underwent political violence and economic stagnation. Today, the economy of West Bengal is the sixth-largest state economy in India with a gross state domestic product (GSDP) of ₹12.54 lakh crore (US$180 billion), and has the country's 23rd-highest GSDP per capita of ₹115,748 (US$1,600).

West Bengal also has the 28th-highest ranking among Indian states in human development index, with the index value being less than that of India. The state government debt of ₹4.0 lakh crore (US$56 billion), or 32.6% of GSDP, is fifth highest India, but has dropped from 40.65% since 2010–11. There is moderate unemployment. West Bengal has two World Heritage sites and ranks as the seventh-most visited tourist destination in India.

Table 4.1:General Information of West Bengal

Total population	91,347,736
Male	46,927,3 89
Female	44,420,347
Population Growth	13.93%
Sex ratio	947
Population density	1029/ km
Literacy rate	77.08%
Male	82.67%
Female	71.16%
No of Sub divisions	66
No of blocks	341
No of Gram Panchayats	3354
Inhabited Villages	37945
Area (Sq. Km)	88752
% of Electrified Villages	83.6
Source: Office of the Register General and Census Commissioner 2011 Ministry of Home Affairs, GOI.	

Geography and Climate of West Bengal

West Bengal is on the eastern bottleneck of India, stretching from the Himalayas in the north to the Bay of Bengal in the south. The state has a total area of 88,752 square kilometers (34,267 sq. mi). The Darjeeling Himalayan hill region in the northern extreme of the state is a part of the eastern Himalayas mountain range. In this region is Sandakfu, which, at 3,636 m (11,929 ft.), is the highest peak in the state. The narrow Terai region separates the hills from the North Bengal plains, which in turn transitions into the Ganges delta towards the south. The Rarh region intervenes between the Ganges delta in the east and the western plateau and high lands. A small coastal region is in the extreme south, while the Sundarbans mangrove forests form a geographical landmark at the Ganges delta.

The main river in West Bengal is the Ganges, which divides into two branches. One branch enters Bangladesh as the Padma, while the other flows through West Bengal as the Bhagirathi River and Hooghly River. The Farakka barrage over the Ganges feeds the Hooghly branch of the river by a feeder canal. Its has been a source of lingering dispute between India and Bangladesh. The Teesta, Torsa, Jaldhaka and Mahananda rivers are in the northern hilly region. The western plateau region has rivers like the Damodar, Ajay and Kangsabati. The Ganges delta and the Sundarbans area have numerous rivers and creeks. Pollution of the Ganges from indiscriminate waste dumped into the river is a major problem. Damodar, another tributary of the Ganges and once known as the "Sorrow of Bengal" (due to its frequent floods), has several dams under the Damodar Valley Project. At least nine districts in the state suffer from arsenic contamination of groundwater, and as of 2017 an estimated 1.04 crore people were afflicted by arsenic poisoning.

West Bengal's climate varies from tropical savanna in the southern portions to humid subtropical in the north. The main seasons are summer, the rainy season, a short autumn and winter. While the summer in the delta region is noted for excessive humidity, the western highlands experience a dry summer like northern India. The highest daytime temperatures range from 38 °C (100 °F) to 45 °C (113 °F). At night, a cool southerly breeze carries moisture from the Bay of Bengal. In early summer, brief squalls and thunderstorms known as *Kalbaisakhi*, or Nor'westers, often occur.

West Bengal receives the Bay of Bengal branch of the Indian Ocean monsoon that moves in a southeast to northwest direction. Monsoons bring rain to the whole state from June to September. Heavy rainfall of above 250 centimeters (98 in) is observed in the Darjeeling, Jalpaiguri, and Cooch Behar district. During the arrival of the monsoons, low pressure in the Bay of Bengal region often leads to the formation of storms in the coastal areas. Winter (December–January) is mild over the plains with average minimum temperatures of 15 °C (59 °F). A cold and dry northern wind blows in the winter, substantially lowering the humidity level. The Darjeeling Himalayan Hill region experiences a harsh winter, with occasional snowfall.

Agriculture of West Bengal

West Bengal is predominantly an agrarian State. Comprising of only 2.7% of India's geographical area, it supports nearly 8% of its population. There are 71.23 lakh farm families of whom 96% are small and marginal farmers. The average size of land holding is only 0.77 ha. However, the State is bestowed with diverse natural resources and varied agro-climatic conditions which support cultivation of a wide range of crops.

West Bengal ranks first in paddy and vegetable production in the country. The net cropped area is 52.05 lakh ha which comprises 68% of the geographical area and 92% of arable land. The cropping intensity is 184%. However, as the State is located in the humid tropic and the Bay of Bengal is close by, it has to often face vagaries of nature like flood, cyclone, hailstorm etc. Though the State has a surplus production of rice, vegetables and potato, a huge gap exists between the requirement and production of pulses, oilseeds and maize. Deterioration of soil health due to imbalance in the use of chemical fertilizers, paucity of suitable improved varieties of seed, inadequate farm mechanization, unorganized marketing structure etc. are major challenges to agricultural growth.

Table 4.2: Land Use Pattern of West Bengal

Geographical Area (MHA)	8.86
Net Cropped Area (MHA)	5.46
Gross Cropped Area (MHA)	9.23

Irrigated Area (MHA)	4.67
Cropping Intensity (%)	165.34
Forest Area (%)	13.72
Net Agricultural Use (%)	18.66
Barren and Uncultivable (%)	0.35
Permanent pastures and Grazing (%)	0.09
Trees and Groves (%)	0.88
Cultivable wasteland (%)	0.53
Fallow land (%)	0.34

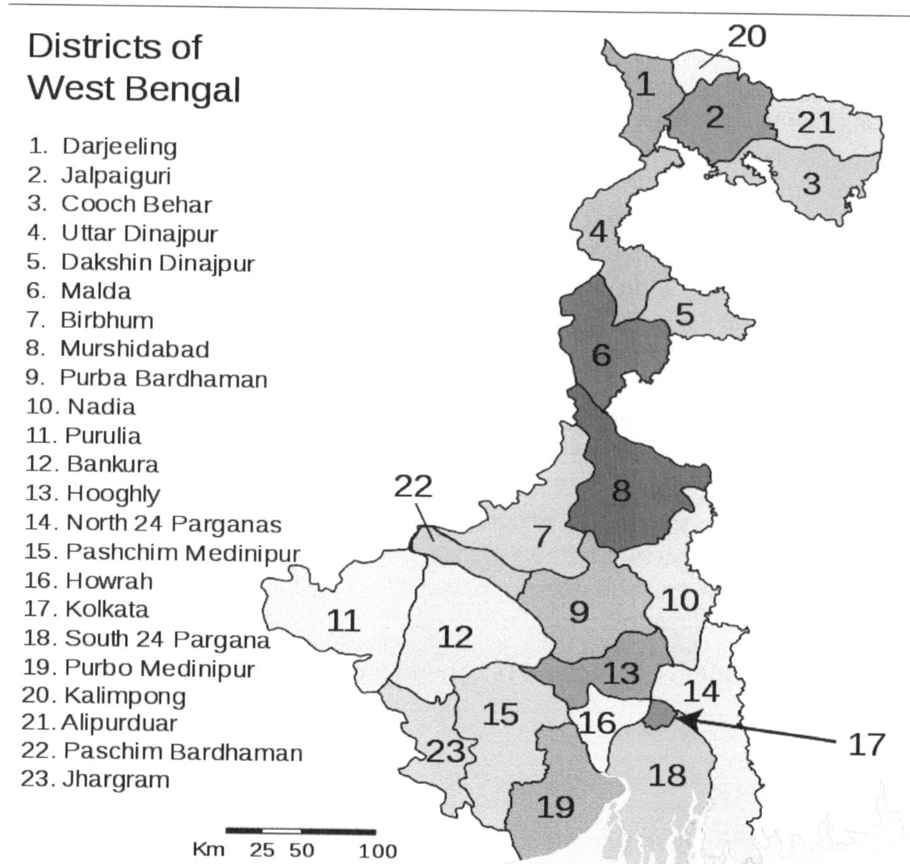

Map-1: Political Map of West Bengal

Profile of Nadia District

The district of Nadia is situated in the heart of the Bengal delta held within the arms of the Ganga, namely, the Bhagirathi on the West and Mathabhanga on the North. The entire district lies in the alluvial plain of the Ganga and its tributaries. Most districts in West Bengal take their name from the headquarters station of the district, but Nadia district takes its name not from Krishnagar, the headquarter but from Nadia or Nabadwip hallowed by the memory of Lord Shri Chaitanya

Mahaprabhu who was born here on 18th February, 1486. The British district of Nadia was formed in 1787. The present district of Nadia after partition was formed by Notification No.545 GA dated 23rd February, 1948.

The district of Nadia is bounded on the north and north west by the district of Murshidabad. On the south east and east it is bounded by the Republic of Bangladesh. In the south and south east, the district is bounded by the district of North 24 Parganas. The shape of the district is irregular, lying North to South. The district is about 46ft. above the mean sea level and the tropic of cancer divides the district into two parts.

For administrative convenience Nadia district is divided into four Subdivisions – Krishnagar Sadar, Ranaghat. Kalyani and Tehatta. The district has an area of 3927 sq. kms having a population of 46, 04,827 as per Census 2001. Out of that SC & ST population are 13,65,985 and 1,13891 respectively. The district has 17 Panchayat Samities consist of 187 Gram Panchayats and 8 Municipalities. Total number of Police Stations in the district is 19. The density of population in this district is 1173 persons per sq. km. Nadia district has 950 females as against 1000 males. The majority of the people of the district speak Bengali followed by Hindi, Santali and other. Religion wise about 73.75% are Hindus and 25.42% are Muslims. In the district of Nadia, the percentage of literacy by sex is 72.30 (Male) & 59.60 (Female) as per Census 2001. The important rivers of the district are Bhagirathi, Churni, Mathabhanga, Ichamati and Jalangi. Krishnanagar is the most important town of this district, as it is the district headquarters of Nadia. It is on the bank of river Jalangi. The palace of Nadiaraj (Rajbari) built here is a prominent place of tourist attraction though the remnants of the past glory have been eroded and only a dilapidated structure of the exquisite places with carving on its inner walls exists today. Krishnagar is famous for world class clay models & terracotta works. It can claim a proud tradition for the sweets, Sarpuria & Sarbhaja.

Table 4.3: General Information of Nadia District

Total population	5,167,600
Male	2,653,768
Female	2,513,832
Population Growth	12.22%
Sex ratio(per 1000)	947
Population density	1300/ km2
Literates Rate	74.97%
Male	78.75%
Female	70.98%
No of Sub divisions	5
No of Development blocks	17
No of Gram Panchayats	187
Inhabited Villages	2639
Area (Sq. Km)	3,927

Source: Office of the Register General and Census Commissioner 2011 Ministry of Home Affairs, GOI.

Climate of Nadia District

The climate in Nadia district is almost the same as in other parts of West Bengal except the Northern cold region. Cold weather starts from the middle of November and continued to the end of February. Hot summer, high humidity and well distributed rainfall during the South-West Monsoon season are the main features of climate. During March and April, the weather becomes increasingly hot during the day and temperature is frequently very high. South-West Monsoon starts from the beginning of June and last to the end of September.

Table 4.4: The maximum and minimum temperature of the district for the preceding 5 years in degree Celsius

Year	*Maximum temperature (°C)*	*Minimum temperature (°C)*
2007	39	8
2008	37	9
2009	41	8
2010	41	7
2011	37	5
(*Source*: District Statistical Hand Book, Nadia, 2010-11)		

The level of Annual rainfall in the district is pretty high barring rare exceptions.

Table 4.5: The status of Annual Rainfall during the period of 2007-2011

Year	*Annual Rainfall (in mm.)*
2007	1612
2008	1436
2009	1209
2010	940
2011	1488

The level of Annual rainfall in the district is pretty high barring rare exceptions.

Soil & Cropping Pattern

The soil type in Nadia district is new alluvium. Almost all important crops are profitably grown in this district. The economy of the district is depending upon agriculture. The soil is suitable for paddy cultivation. In many parts of the district farmers cultivate wheat, pulses and jute. The low lying tract in the district which is known as Kalantar is also good for cultivation of aman rice.

Distribution of Operational Holdings Over Size–Classes

(Area in hectares)

	SIZE - CLASS												
	Marginal		Small		Semi–medium		Medium		Large		Total		Average size of Holdings (hect.)
Year	No. of Holdings	Area of Holdings	No. of Holdings	Area of Holdings	No. of Holdings	Area of Holdings	No. of Holdings	Area of Holdings	No. of Holdings	Area of Holdings	No. of Holdings	Area of Holdings	
(1)	(2)	(3)	(4)	(5)	(6)	(7)	(8)	(9)	(10)	(11)	(12)	(13)	(14)
2000–01	310374	180534	76386	130547	12658	34159	1159	5858	19	870	400596	351968	0.88
2005–06	329150	182763	75752	128679	12443	33910	880	4822	13	415	418238	350589	0.84

Note : Marginal = Below 1.0 hectare
Small = 1.0 hectare and above but less than 2.0 hectares
Semi–medium = 2.0 hectares and above but less than 4.0 hectares
Medium = 4.0 hectares and above but less than 10.0 hectares
Large = 10.0 hectares and above. It includes mostly institutional holdings

(Source : District Statistical Hand Book, Nadia, 2010-11)

Land & Land-use Pattern

Nadia district comprises of very good fertile land suitable for agricultural crop cultivation. The important crops cultivated in the district are paddy, wheat, jute, pulses. Vegetables are also produced in the district in very good quality. The following table shows the land use pattern in the district.

General Land use pattern of Nadia district

(Area in thousand hectares)

Year	Reporting Area	Forest Area	Area under Non-agri-cultural use	Barren & uncul-turable land	Permanent pastures & other grazing land	Land under misc. tree groves not included in Net area sown	Culturable waste land	Fallow land other than current fallow	Current fallow	Net area sown
(1)	(2)	(2)	(3)	(4)	(5)	(6)	(7)	(8)	(9)	(10)
2006-07	390.66	1.22	86.52	0.17	0.07	4.40	0.70	0.68	6.90	290.00
2007-08	390.67	1.22	88.36	0.12	0.04	4.76	0.82	0.27	5.83	289.25
2008-09	390.66	1.22	90.14	0.21	0.11	3.78	0.90	0.08	5.05	289.17
2009-10	390.66	1.22	88.94	0.24	0.05	3.91	0.79	0.37	4.40	290.74
2010-11	390.66	1.22	90.22	0.06	0.06	3.73	0.63	0.11	4.18	290.45

(Source : District Statistical Hand Book, Nadia, 2010-11)

Agriculture

Nadia district remains evergreen with seasonal field crops throughout the year and almost all important crops are grown profitably in the district. The economy of the district mainly depends on agriculture.

Irrigation

River is the main source of irrigation in the district. Lift irrigation and tube wells are also installed for watering the field under crop. The following table shows various sources of irrigation for the period from 2006-2011.

Sources of Irrigation in Nadia district

(in number)

Year	Tank	HDTW	MDTW	LDTW	STW+	RLI	ODW	Others
(1)	(2)	(3)	(4)	(5)	(6)	(7)	(8)	(9)
2006-07	---	680	33	293	826	322	---	---
2007-08	---	680	33	293	826	322	---	---
2008-09	---	613	32	222	807*	293	---	---
2009-10	---	609	31	216	502*	293	---	1
2010-11	---	612	31	228	588*	293	---	1

(Source : District Statistical Hand Book, Nadia, 2010-11)

HDTW = High Capacity Deep Tubewell
MDTW = Middle Capacity Deep Tubewell
LDTW = Low Capacity Deep Tubewell
STW = Shallow Tubewell
RLI = River Lift Irrigation
ODW = Open Dug Well

Industry

As said earlier, Nadia is a basically an agricultural oriented district. No major industries are found in the district. The district is rich with paddy, jute, mango, papaya, lichee, tomato, ginger, garlic and other vegetables. Many skilled artisans are engaged in various handicrafts like clay dolls and modeling, brass and bell metal, needle work, wood carvings, cane and bamboo crafts etc. Manufacturing of cycle parts, production of yarns, re-rolling of steel rods etc. are the important industry found in the district. Power loom and cotton weaving units are found in the district.

Shantipur weaved clothes are very famous and industries like brass and bell metal work, pottery clay modeling and mat are found in this district.

The important industries in the district are pottery and clay modeling, brass and bell metal industry, engineering industry, automobile industry, food processing industry, cane and bamboo products, silver ornaments manufacturing, sola (thermocol) crafts, hosiery product, silk needing, vegetable processing, jute processing, oil processing, agro service centers and ice factories etc.

Trade and Commerce

Ranaghat is the important commercial centre in Nadia district. It is the center of jute industry and jute is the major item available here for export. Among other exportable items gram and pulses, vegetables, potatoes, sugar, and molasses are included. The major export destination is Kolkata.

The imported items in the district are rice, coal, salt and kerosene. Clay dolls and models manufactured at Krishnanagar are world famous. The demand of these clay dolls and models are not only restricted to Indian market but are also well acclaimed internationally.

Transport

The main transport link in the district is between District Head Quarter Krishnanagar and Sub-divisional head quarter Ranaghat. The Eastern Railway has provided railway junction in both the places. The road connecting East to West starts from Krishnanagar to Nabadwip. It also extends towards South to Shantipur. Roads in the district maintained by Government of West Bengal, PWD and local authorities. Municipalities in the district also maintained certain important road links in the district.

Electricity & Power:

Nadia district consists of 10 statuary towns and 1307 villages. Total power consumption in the district by various schemes is given below.

Consumption of Electricity by Different Sectors in Nadia district

(Thousand K.W.H.)

Year	Domestic	Commercial	Industrial	Public Lighting	Agricultural irrigation & dewatering	Public Water Works & Sewerage Pump	Railway Traction & Non-Traction	Miscella -neous	Total
(1)	(2)	(3)	(4)	(5)	(6)	(7)	(8)	(9)	(10)
2006-07	271640	74080	163340	6050	176610	7940	23490	-	723150
2007-08	273185	77758	166197	6312	137564	8475	28028	-	697519
2008-09	295957	61814	59992	7619	105096	7976	1230	9445	549129
2009-10	323546	68697	62486	8641	112221	7060	1419	6160	590230
2010-11	370890	84607	62985	10361	128529	8083	1687	2730	669872

(Source : District Statistical Hand Book, Nadia, 2010-11)

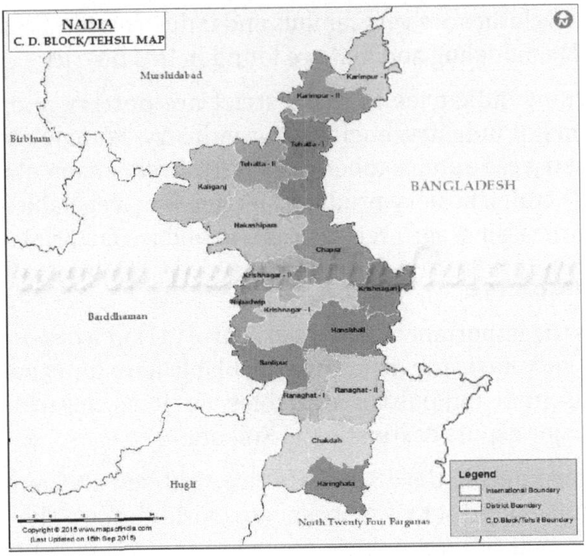

Map-2: Map of Nadia District

Profile of Haringhata Block

Haringhata is a town in Kalyani subdivision of Nadia district in the state of west Bengal of India.

Location

Haringhata is located at 22°95'N 88°57'E. It has an average elevation of 10 meters. Haringhata community development block has an area of 168.59 Km^2.

Language

Bengali is the local language in these areas.

Table 4.6 General Information about Haringhata block

Block area	170.32 sq. Km
No. of gram panchayats	10
No. of gram sabhas	129
No. of Mouza	87
No. of police stations	1
Elevation	13 m (43 ft.)
Total Population (2011)	231,068
Density	1,400/km2(3,500/sq. mi)
Languages	Bengali, English
Total literates	171,144 (82.15%)

Geography of Haringhata Block

Haringhata Block is bounded by Chakdah Block in the north and west, Bangaon and Gaighata Blocks, in North 24 Parganas district, in the east, Amdanga, Habra I and Habra II Blocks, in North 24 Parganas district, in the south. Nadia district is mostly alluvial plains lying to the east of Hooghly River, locally known as Bhagirathi. The alluvial plains are cut across by such distributaries as Jalangi, Churni and Ichhamati. With these rivers getting silted up, floods are a recurring feature. Haringhata Block has an area of 170.32 km2. It has 1 panchayat samity, 10 gram panchayats, 176 gram sansads (village councils), 87 mouzas and 82 inhabited villages. police station serves this block. Headquarters of this Block is at Subarnapur. It is located 56 km from Krishnanagar, the district headquarters.

Haringhata Block/ Panchayat Samiti are: Birohi 1, Birohi 2, Fatepur, Haringhata 1, Harinhgata 2, Kastodanga 1, Kastodanga 2, Mollabelia, Nagarukhra 1 and Nagarukhra 2.

Demographics

Population

As per the 2011 Census of India, Haringhata Block had a total population of 231,068, of which 207,459 were rural and 23,609 were urban. There were 118,709 (51% males and 112,359 (49%) females. Scheduled Castes numbered 73,755 (31.92%) and Scheduled Tribes numbered 11,218 (4.85%).

As per the 2001 census, Haringhata block had a total population 207,867, out of which 108,011 were males and 99,856 were females. Haringhata block registered a population growth of 16.92 per cent during the 1991 2001 decade. Decadal growth for the district was 19.51 per cent. Decadal growth in West Bengal was 17.84 per cent.

There are four census towns in Haringhata Block (2011 census figures in brackets): Haringhata (3,989), Simhat (6,949), Subarnapur (5,759) and Digha (6,916).

Large villages (with 4,000+ populations) in Haringhata Block were (2011 census figures in brackets): Chandirampur (4,740), Narayanpur (4,548), Uttar Datta Para (5,430), Uttar Brahmapur (5,371), Subuddipur (6,051), Sekendarpur (7,917), Kathdanga (7,417), Mollabalia (5,889), Ganguria (4,291), Fatepur (5,093), Barasat (4,046),Ukrah (13,548), Goaldob (4,883) and Dighalgram (5,390).

Literacy

As per the 2011 census, the total number of literates in Haringhata CD Block was 171,144 (82.15% of the population over 6 years) out of which males numbered 93,313 (87.11% of the male population over 6 years) and females numbered 77,831 (76.90% of the female population over 6 years). The gender disparity (the difference between female and male literacy rates) was 10.21%.

Language and religion

Bengali is the local language in these areas.

In the 2011 census, Hindus numbered 164,263 and formed 71.09% of the population in Haringhata Block. Muslims numbered 65,148 and formed 28.19% of the population. Christians numbered 460 and formed 0.20% of the population. Others numbered 1,197 and formed 0.52% of the population.

In the 2001 census, Hindus numbered 151,379 and formed 72.79% of the population of Haringhata CD Block. Muslims numbered 55,885 and formed 26.97% of the population. In the 1991 census, Hindus numbered 130,125 and formed 73.19% of the population of Haringhata CD Block. Muslims numbered 47,284 and formed 26.60% of the population.

Rural poverty

The District Human Development Report for Nadia has provided a CD Block wise data table for Modified Human Vulnerability Index of the district. Haringhata CD Block registered 31.59 on the MHPI scale. The CD Block wise mean MHVI was estimated at 33.92. A total of 8 out of the 17 CD Blocks in Nadia district were found to be severely deprived when measured against the CD Block mean MHVI Karimpur I and Karimpur II (under Tehatta subdivision), Kaliganj, Nakashipara, Chapra, Krishnanagar I and Nabadwip (under Krishnanagar Sadar subdivision) and Santipur (under Ranaghat subdivision) appear to be backward.

Economy

Livelihood

In Haringhata CD Block in 2011, amongst the class of total workers, cultivators formed 18.42%, agricultural labours 35.60, household industry workers 3.79% and other workers 42.19%.

The southern part of Nadia district starting from Krishnanagar I down to Chakdah and Haringhata has some urban pockets specializing in either manufacturing or service related economic activity and has reflected a comparatively higher concentration of population but the urban population has generally stagnated. Nadia district still has a large chunk of people living in the rural areas.

Infrastructure

There are 82 inhabited villages in Haringhata Block. 100% villages have power supply and drinking water supply. 13 Villages (15.85%) have post offices. All 82 villages (100%) have telephones (including landlines, public call offices and mobile phones). 64 villages (78.05%) have a pucca approach road and 30 villages (38.59%) have transport communication (includes bus service, rail facility and navigable waterways). 6 villages (7.32%) have agricultural credit societies and 8 villages (9.76%) have banks. It should, however, be noted that although 100%

villages in Nadia district had power supply in 2011, a survey in 2007 08 revealed that less than 50% of households had electricity connection. In rural areas of the country, the tube well was for many years considered to be the provider of safe drinking water, but with arsenic contamination of ground water claiming public attention it is no longer so. Piped water supply is still a distant dream. In 2007 08, the availability of piped drinking water in Nadia district was as low as 8.6%, well below the state average of around 20%.

Agriculture

Although the Bargadari Act of 1950 recognized the rights of bargadars to a higher share of crops from the land that they tilled, it was not implemented fully. Large tracts, beyond the prescribed limit of land ceiling, remained with the rich landlords. From 1977 onwards major land reforms took place in West Bengal. Land in excess of land ceiling was acquired and distributed amongst thepeasants. Following land reforms land ownership pattern has undergone transformation. In 2013 14, persons engaged in agriculture in Haringhata Block could be classified as follows: bargadars 2.44%, patta (document) holders 7.95%, small farmers (possessing land between 1 and 2 hectares) 6.18%, marginal farmers (possessing land up to 1 hectare) 33.53% and agricultural labours 49.09%. As the proportion of agricultural labours is very high, the real wage in the agricultural sector has been a matter of concern.

Haringhata CD Block had 207 fertilizer depots, 40 seed store and 44 fair price shops in 2013-14.

In 2013-14, Haringhata CD Block produced 72,866 tonnes of Aman paddy, the main winter crop from 24,942 hectares, 20,348 tonnes of Boro paddy (spring crop) from 5,720 hectares, 5,413 tonnes of Aus paddy (summer crop) from 2,013 hectares, 4,317 tonnes of wheat from 1,315 hectares, 42,799 tonnes of jute from 2,383 hectares and 16,616 tonnes of potatoes from 536 hectares. It also produced pulses and oilseeds.

In 2013-14, the total area irrigated in Haringhata Block was 967 hectares, out of which 89 hectares were irrigated by river lift irrigation, 830 hectares by deep tube wells and 48 hectares by shallow tube wells.

Haringhata Dairy Farm

In order to supply milk to the urban population in West Bengal, the first dairy plant was installed in 1950 at Haringhata by the Directorate of Dairy Development, under the aegis of Animal Resources Development Department, Government of West Bengal. It started with a production of 1.5 thousand litres of bottled milk per day in 1950 51. With a network of two dairy plants, seven milk colony units and 19 Milk Collection cum Chilling stations named as —Greater Calcutta Milk Supply Scheme || the production had gradually gone up to 210 thousand litres per day by 1980s. (Directorate of Dairy Development, under the aegis of Animal Resources Development Department, Govt. of West Bengal)

Banking

In 2013-14, Haringhata Block had offices of 12 commercial banks and 2 gramin banks.

Transport

Haringhata CD Block has 4 originating/ terminating bus services. The nearest railway station Kanchrapara is 16 km and Kalyani Railway Station is 19 km from Block headquarters.

NH 12 (old number NH-34) passes through this block. SH1, running from Bangaon (in North 24 Parganas district) to Kulpi (in South 24 Parganas district) passes through this Block.

Education

In 2013 14, Haringhata Block had 120 primary schools with 9,251 students, 11 middle schools with 866 students, 3 high schools with 1,819 students and 18 higher secondary schools with 23,167 students. Haringhata Block had 1 general college with 3,233 students, 2 technical/ professional institutions with 1,880 students and 453 institutions for special and no normal education with 10,874 students. In Haringhata Block, amongst the 82 inhabited villages, 4 villages did not have any school, 21 had more than 1 primary school, 16 had at least 1 primary and one middle school and 14 had 1 middle school and 1 secondary school.

Healthcare

In 2014, Haringhata Block had 1 rural hospital, 4 primary health centres and 2 private nursing homes with total 96 beds and 9 doctors (excluding private bodies). It had 30 family welfare sub centres. 4,907 patients were treated indoor and 357,557 patients were treated outdoor in the hospitals, health centres and sub centres of the Block.

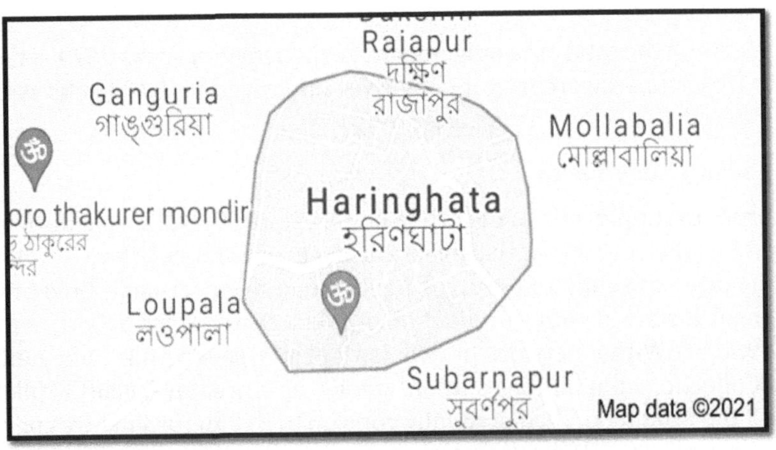

Map-3: Map of Haringhata Block

Haringhata Rural Hospital at Mohanpur has 25 beds. There are primary health centres at Birohi (6 beds), Kasthodanga (6 beds) and Nagarukhra (10 beds). Haringhata Block is one of the areas of Nadia district where ground water is affected by high level of arsenic contamination. The WHO guideline for arsenic in drinking water is 10 mg/ litre, and the Indian Standard value is 50 mg/ litre. All the 17 blocks of Nadia district have arsenic contamination above this level. The maximum concentration in Haringhata Block is 865 mg/litre.

Profile of Village

The area of investigation is located in West Bengal state, in eastern part of India. This state has as unique socio-cultural and ecological background, which influence the living standard of the inhabitant in many ways. The area of the investigation belongs to Chak Birohi, Kantabele, Ghoragacha, Mitrapur and Kashtodanga of Nadia district.

The villages are selected after careful consideration of status and dimension of productivity, income, livelihood, ecology and wage scenario of the above mentioned villages. The pilot study shows that the selected villages, for the last two decades, have been passing through agricultural modernization along with some implicit and explicit contradiction as well as conciliation. Due to pandemic situation, the researcher has been constraint to go for non random sampling method, especially snowballing through referencing and cross referencing. The data have been collected as per the convenience of the target respondents looking into their ease of communication and least of disruption of their ongoing farm operations. Here under the profile of the selected villages has been systematically presented to have a reference of changes and status, movement and undulation along and across the terrains of ecology, economy and chronology.

Profile of Mitrapur Village

According to Census 2011 information the location code or village code of Mitrapur village is 322352. Mitrapur village is located in Haringhata subdivision of Nadia district in West Bengal, India. It is situated 36.3km away from sub-district headquarter Subarnapur (tehsildar office) and 53.8km away from district headquarter Krishnanagar. As per 2009 stats, Mollabelia is the gram panchayat of Mitrapur village.

The total geographical area of village is 298.02 hectares. Mitrapur has a total population of 3,048 peoples, out of which male population is 1,593 while female population is 1,455. Literacy rate of mitrapur village is 68.08% out of which 72.44% males and 63.30% females are literate. There are about 707 houses in mitrapur village. Pincode of mitrapur village locality is 741257.

Particulars	*Value*
Sex Ratio	1043
Total Household	236

Household Owing Land		34
House Ownership	Owned	230
	Rented	1
	Any other	5
Land Owned	Total unirrigated land (in hectares)	11.67
	Total irrigated land (in hectares)	352.33
	Total other irrigated land (in hectares)	9.33
Households with Highest Earning member Income as	Less than Rs. 5,000	190
	Between Rs. 5,000 and Rs 10,000	46
	Rs. 10,000 or more	0
Main sources of household income	Cultivation	36
	Manual Casual Labour	127
	Part-time or Full Time Domestic Service	2
	Foraging Rag Picking	0
	Non-agricultural Own Account Enterprise	0
Total population		993
SC population		75
ST population		0
Male		486
Female		507
[Source: https://secc.gov.in/homePageLgd.htm]		

Kastodanga Village

Kastodanga is a village in Haringhata Block in Nadia District of West Bengal State, India. The village Kastodanga falls under the gram panchayat Kastodanga-I. It belongs to Presidency Division. It is located 59 KM towards South from District headquarters Krishnanagar, Nadia. 7 KM from Haringhata. 60 KM from State capital Kolkata. Kastodanga-I Pin code is 741257 and postal head office is Nagarukhra.

Table:4.8. Cohort study of Kastodanga village

COHORT				
SL No.	***Particulars/Indicators***	***Years***		
		1990 - 2000	**2000 - 2010**	**2010 - 2020**
1	Crop Combination	Rice, jute, mustard, brinjal, parwal, chickpea, lentil	Rice, jute, cauliflower, chilli, bitter gourd, flat beans	Flat beans, boro rice(cultivation increased), jute cultivation decreased and has been replaced by vegetables.

2	Rice Yield (per bigha)	25 mon 1 mon= 40 kg	21-22 mon	18-20 mon
3	Cost of Cultivation of Rice (Rs./bigha)	4000	5500	10000
4	Soil Health	Very good	medium	poor
5	Water Availability	Bhomra Bil	Bhomra Bil	Shallow
6	Tree and crop biodiversity	Tree: Shishu, Aswatta,Banyan Rice variety: Jaya,IR50,Ratna, Lal Swarna	Tree: Eucalyptus, Mahogany Rice variety: Ranjit, Pankaj	Tree: Lombu, Kadam Rice variety: Satabdi, Gosai
7	Application of Organic Manure (cow dung)	More	Medium	Very less
8	Livestock Count (cows)	1500	600	30
9	No. of ponds	112	122	135
10	Rainfall (Rainy Days)	High (4-5 months)	Medium (3-4 months)	Less (2 months)
11	Income/Net Return (Rs. /month/capita)	3000	5000	8000
12	Expenditure (Rs./ month)	1000	4000	6500
13	Well-being (Score) (out of 10)	5	7	8
(***Source:*** onefivenine.com and Bharat Chandra Bachar, a farmer of Kastodanga)				

Kastodanga-I is surrounded by Habra-I Block towards South, Chakdah Block towards West, Bangaon Block towards East, Gaighata Block towards East. This Place is in the border of the Nadia District and 24 Parganas North District. 24 Parganas North District Habra-I is South towards this place.Bengali is the Local Language here.Total Population of the village is 18019 approximately, where male population is about 9817 and female population is about 8502. Public bus service is available within 5 km distance and railway station within 17 km distance.

Profile of Chak Birahi village

Particulars		*Value*
Sex Ratio		1016
Total Household		63
Household Owing Land		10
House Ownership	Owned	74
	Rented	0
	Any other	16
Land Owned	Total unirrigated land (in hectares)	1.34
	Total irrigated land (in hectares)	7.88
	Total other irrigated land (in hectares)	0.16

Particulars		*Value*
Households with Highest Earning member Income as	Less than Rs. 5,000	60
	Between Rs. 5,000 and Rs 10,000	3
	Rs. 10,000 or more	0
Main sources of household income	Cultivation	6
	Manual Casual Labour	53
	Part-time or Full Time Domestic Service	2
	Foraging Rag Picking	0
	Non-agricultural Own Account Enterprise	1
Total population		260
SC population		15
ST population		0
Male		129
Female		131
Distribution of population by age	0-5 Yrs	27
	6-15 Yrs	48
	16-25 Yrs	47
	26-35 Yrs	63
	36-45 Yrs	30
	46-55 Yrs	20
	56-65 Yrs	14
	66-75 Yrs	9
	Above 76 Yrs	2
Population by marital status	Never Married	118
	Currently Married	128
	Widowed	12
	Separated	2
	Divorced	0
Population by highest education level completed	Illiterate	115
	Below Primary	30
	Primary	32
	Middle	65
	Secondary	10
	Higher Secondary	5
	Graduate or Above	3
	Other	0
[***Source:*** https://secc.gov.in/homePageLgd.htm]		

Profile of Ghoragacha village:

Particulars		Value
Sex Ratio		921
Total Household		412
Household Owing Land		106
House Ownership	Owned	412
	Rented	0
	Any other	0
Land Owned	Total unirrigated land (in hectares)	59.39
	Total irrigated land (in hectares)	68.31
	Total other irrigated land (in hectares)	13.65
Households with Highest Earning member Income as	Less than Rs. 5,000	280
	Between Rs. 5,000 and Rs 10,000	96
	Rs. 10,000 or more	36
Main sources of household income	Cultivation	48
	Manual Casual Labour	285
	Part-time or Full Time Domestic Service	0
	Foraging Rag Picking	0
	Non-agricultural Own Account Enterprise	14
Total population		1883
SC population		256
ST population		0
Male		980
Female		903
Distribution of population by age	0-5 Yrs	145
	6-15 Yrs	355
	16-25 Yrs	432
	26-35 Yrs	343
	36-45 Yrs	292
	46-55 Yrs	158
	56-65 Yrs	82
	66-75 Yrs	54
	Above 76 Yrs	22

Particulars		Value
Population by marital status	Never Married	785
	Currently Married	998
	Widowed	88
	Separated	10
	Divorced	0
Population by highest education level completed	Illiterate	606
	Below Primary	116
	Primary	334
	Middle	582
	Secondary	115
	Higher Secondary	56
	Graduate or Above	73
	Other	1
[***Source***: https://secc.gov.in/homePageLgd.htm]		

Profile of Kata Belia village:

Particulars		Value
Sex Ratio		1043
Total Household		236
Household Owing Land		34
House Ownership	Owned	230
	Rented	1
	Any other	5
Land Owned	Total unirrigated land (in hectares)	11.67
	Total irrigated land (in hectares)	352.33
	Total other irrigated land (in hectares)	9.33
Households with Highest Earning member Income as	Less than Rs. 5,000	190
	Between Rs. 5,000 and Rs 10,000	46
	Rs. 10,000 or more	0
Main sources of household income	Cultivation	36
	Manual Casual Labour	127
	Part-time or Full Time Domestic Service	2
	Foraging Rag Picking	0
	Non-agricultural Own Account Enterprise	0
Total population		993
SC population		75
ST population		0
Male		486
Female		507

Particulars		*Value*
Distribution of population by age	0-5 Yrs	83
	6-15 Yrs	183
	16-25 Yrs	220
	26-35 Yrs	183
	36-45 Yrs	170
	46-55 Yrs	61
	56-65 Yrs	62
	66-75 Yrs	23
	Above 76 Yrs	8
Population by marital status	Never Married	428
	Currently Married	507
	Widowed	51
	Separated	4
	Divorced	3
Population by highest education level completed	Illiterate	268
	Below Primary	51
	Primary	441
	Middle	148
	Secondary	52
	Higher Secondary	25
	Graduate or Above	8
	Other	0
[***Source***: https://secc.gov.in/homePageLgd.htm]		

Chapter 5

RESEARCH METHODOLOGY

Research methodology is a detailed plan of investigation and the blue print of procedure for carrying out the research. In this chapter, discussion on the methodology has been made to understand the concepts, methods and techniques, which are utilized to design the study, collect information, analyzing data and interpreting the findings for revelation of truth and formulation of theories. The entire chapter has been broken up under following sub-heads for easy understanding:

- Locale of research
- Sampling design
- Pilot study
- Variables and their measurements
- Methods of data collection
- Statistical tools used for analysis of data

Locale of Research

The present study has been conducted in New Alluvial agro-ecological zones of the state West Bengal. Five villages from Haringhata block of Nadia district were selected from the new alluvial agro- ecological zone to conduct the study. The characters and the factors under study have been well discernible to this area

- The researcher's close familiarity with respect to area, people, officials and local dialects
- The ample opportunity to generate relevant data due to the close proximity of the area with the research and extension wing of the State Agricultural Universities;

- The highly cooperative and responsive respondents
- The profuse scope to get relevant information regarding perception on management of water resources, application of nutrients, improving soil organic carbon and problems and prospects of practicing conservation agricultural practices
- Experienced, well versed, venturesome, enthusiast and risk bearing farmers
- Scope of establishing a sustainable and profitable farming opportunities in the area

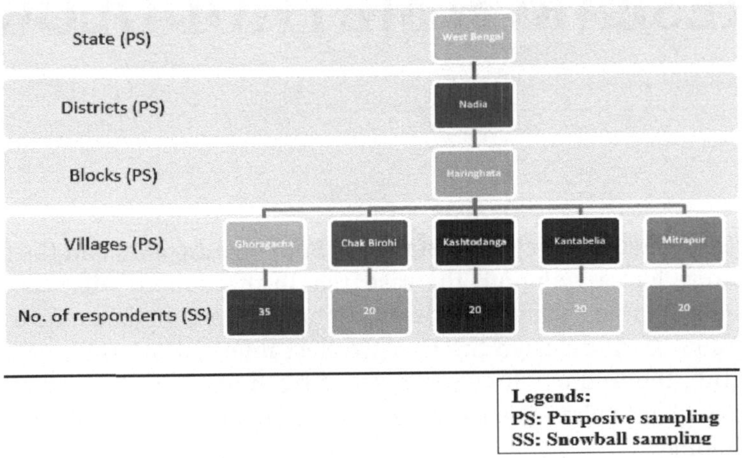

Fig. Schematic Diagram of Sampling Method Followed for the Study

Sampling design

Level	Location	Approach	Justification
State (1)	West Bengal	Purposive	Because of the limitation of time and resources.
District (1)	Nadia	Purposive	High cropping intensity and chances of high natural resource degradation in the region
Block (1)	Haringhata	Purposive	These blocks are the most agriculturally advanced blocks experiencing high cropping intensity and degraded soil health
Villages (5)	Ghoragacha Kantabelia Kashtodanga Chak Birohi Mitrapur	Purposive	Only these were the villages were high profile of modernization
Farmers (105)		Snowball Sampling	Due to COVID 19 pandemic situation, non-random snowball sampling had seemed to be the best suited method

Pilot Study

A pilot study has been conducted in the concerned area before constructing the data collecting devices. In course of this survey, informal discussion has been carried out with some farmers, local leaders, extension agencies and state agricultural officials of the localities. An outline of the socio-economic background of the farmers of the concerned villages, their interest in sustainable agricultural practices, experiences of sustainable farming and management strategies against the adversities helped in preparing and finalizing the interview schedule.

Variables and their Measurement

There are two main categories of variables in the present study-

- Independent variables
- Dependent variables

Independent variables

Independent Variables	*Empirical Measurements*
Age of the respondent (x1)	Chronological age in years
Education (x2)	Through scoring
Family size (x3)	In numbers
Training and capacity building (x4)	In days
Family education score (x5) Total education score of members	
Holding size (x6)	In bigha
Homestead (x7)	In kattha
No. of land fragments (x8)	In numbers
Cultivable land area (x9)	In bigha (0.33 acres)
Water bodies (x10)	In kattha
Weight of brooder fish (x11)	In kg per bigha
Age at injection (x12)	In kilograms
Scientific orientation (x13)	In months
Frequency of injection (x14)	In months
Input management and cost (x15)	In rupees
Other cost input management and cost(x16)	In rupees
Income statements (x17)	In rupees
Productivity (x18)	In value
Area under rice (x19)	In income
Area of vegetable (x20)	In income
Social participation (x21)	In time
Constraint perception(x22)	In frequency

Independent Variables	*Empirical Measurements*
Suggestions for improvement (x23)	In dimensions
Communication through mobile (x24)	In total cost
No. of fragments (x25)	In number
No. of stakeholders (x26)	In number
Family expenditure (x27)	In rupees
Energy consumption (x28)	In per capita
Nutrition and health status (x29)	In quantity

Age (x_1)

Age is one of the most important determinants of social status and social role of the individual. Here age refers to the chronological years completed by the respondents at the time of interview. It has been recorded in integers.

Education (x_2)

Education can be operationalized as the level of formal schooling attained/ literacy acquired by the respondents at the time of interview. Education plays an important role in changing the behavior and outlook of an individual. It provides knowledge and skill to negotiate the dynamics of social life and grant a unique status in the society. Here, level of education has been categorized and scored as per followed:

Categories	*Score*
Illiterate	1
Primary education (upto class 4)	2
Middle school (upto class 8)	3
High school (upto class 10)	4
Intermediate (upto class 12)	5
Graduation & above	6

Family member (x_3)

Family member has been operationalized as the number of members in an individual's family. In this study, only those members have been considered who were staying together and were sharing meal in one chullah.

Training and Capacity holding (x_4)

Trtaining and Capacity holding is measure of farm business. Operationally farm capacity may be defined as a tract of land possessed by an individual for the purpose of growing crops. In the present study, farming land owned by a family is taken as the measure of farm size.

Family Education Score (x_5)

Family Education Score has been operationalized as the total education score divided by total members.

Holding Size (x_6)

Size of holding has been operationalized as the total size of land per bigha

Homestead land (x_7)

Amount of land acquired by the home building and surrounding. This total area was measuired in Cottah.

No. of fragments (x_8)

Land fragments have been perceived in terms of total number of entitlements under the total size of holding. For example- if one ha of land has got three geographical distributions than the number of fragmentation will be three. Here it has been measured by counting the Arithmetic count of pieces of the land (fragments) under the ownership of family.

Cultivable land area (x_9)

Cultivable land area is the piece of land possessed by an individual for the purpose of growing crops. In the present study, actual area under cultivation in bigha was taken as the measure of cultivable land area.

Cropping intensity (x_{10})

Cropping intensity has been operationalized as the proportion of total annual cropped area to the cultivable land area expressed in percentage.

Water bodies(x_{11})

- **Location of water body:** The distance of water body to be covered for monitoring and management from the residential house of fish farmer was considered and ha s been expressed in meter.
- **Depth of water body :** Seasonal data was collected on depth of water and annual average depth of water were calculated and expressed in feet.
- **Age of water body :** The age of each water body(pond) of a fish farmer was taken in years. Then average age of all ponds of the fish farmer was calculated in years.

Weight of brooder fish (x_{12})

It has been quantified in terms of kilogram of brooder fish weight per unit area (cottah) per year.

Age at injection (x_{13})

It has been measured in terms of age of brooder fish expressed through number of month

Frequency of injection (x_{14})

It has been conceived in terms of span of vulnerable growth period expressed in terms of months in which brooder fish undergo administration of hormonal injection.

Input and other management and cost(x_{15})

Costs of all inputs related to fish enterprise have been summated and divided by the area of water body to get the score of the variables, expressed in Rs. Cottah^{-1}yr^{-1}.

Income statements (x_{16})

Total income from fish enterprise of a fish farmer has been calculated and divided by the total area of water body of that fish farmer to get the score of the variable expressed in terms of Rs. Cottah^{-1}yr$^{-1.}$

Productivity (x_{17})

Productivity has been operationalized as the total level of production per unit of input. Crop yields are per acre a measure of productivity.

Area under rice (x_{18})

Area under rice cultivation is defined in terms of total area in which rice is cultivated by a person. The degree of adoption of rice cultivation can be measured from the area under rice.

Area Under Vegetable(x_{19})

Area under vegetable cultivation is defined in terms of total area in which vegetable is cultivated by a person. The degree of adoption of vegetable cultivation can be measured from the area under vegetable.

Social Participation (x_{20})

Scoring was done based on frequency of participation in different social institutions. The summated frequency value has been divided by number of items.

Constraint Perception (x21)

The constraints have been measured by administering a 10 pt. scale on which respondents are expected or made to reach/respond in a classified / catalogued manner and to be transformed into a quantified as well as analyzable quantity.

Perception for Improvement (x22)

Perception for improvement has been measured on 10 point scoring and the scored value has undergone data normalization through z scoring.

Communication through Mobile (x23)

The total frequency of communication in the form of (a) communication received, (b) communication transmitted and (c) communication shared. Then the accrued score has been: $\sqrt[3]{axbxc}$

Number of Fragments (x24)

Land fragments have been perceived in terms of total numbers of entitlements under the total size of holding. For example-if one ha of land has got three geographical distributions than the number of fragmentation will be three. Here it has been measured by counting the Arithmetic count of pieces of the land (fragments) under the ownership of family.

No. of stakeholders (x_{25})

Total number of actors involved in the enterprise.

Family expenditure (x_{26})

Total money value incurred to defray the family expenditure/ family size.

Energy consumption (x_{27})

It is the summation of money value of a) electricity consumption, b) fuel consumption, c) LPG consumption and the accrued value:

Nutrition and health status (x_{28})

Calculated by a)total calorie intake, b) total food intake, c) calculated BMI (Body Mass Index) and the accrued value: .

Dependent variables

Dependent Variables	*Empirical Measurements*
Wages (y1)	Money value of wages per month from farm multiplied by (b) money value from off farm sources per month. The accrued value:
Family Farm Income (y2)	Total money value earned from farming per year under a single family head/ family size
Productivity (y3)	Quintals/ acre of respective commodities (rice, vegetables and fish/ family size).
Income (y4)	Total money value earned per year under a single family head/ family size
Man days (y5)	Total number of working days (man/woman days in a month). This is the mean of the 12 months' total man days in a year expressed in count.
Ecology perception (y6)	The ecology perception as responded and expressed by the respondents over 10 point scoring against a set of selected items done by judges. The score value has undergone data normalization process in the form of z transformation.

Methods Used for Data Collection

Following methods were used for collecting primary data from the respondents:

Preparation of interview schedule

Interview schedule was used as the primary tool for data collection. It may be defined as 'a formal list, a category or inventory and it may be added that it is a counting device used in formal and standardized inquiries, the sole purpose of which is aiding the collection of quantitative and qualitative cross sectional data'.

Comprehensive and detailed interview schedule consisting of structured as well as open ended questions was developed for collecting relevant information from the respondents.

Pre-testing of Interview Schedule

The structured interview schedule was pre tested on a non-sampled area and necessary corrections were made. The interview schedule was finalized after making necessary modification, deletion and addition based on pre testing recommendations.

Statistical Analysis and interpretation of Data (Analytic tools)

The data collected in the interview schedule have been processed and analyzed in accordance with the outline laid down for the purpose at the time of developing the research plan. Processing implies editing, coding, classification and tabulation of collected data. The statistical techniques and tools used in the present study are:

Mean

It is a measure of central tendency (or statistical averages) tells us the point about which items have a tendency to cluster. Such a measure is considered as the most representative figure for the entire mass of data. Measure of central tendency is also known as statistical average. Mean, median and mode are the most popular averages. Mean, also known as arithmetic average, is the most common measure of central tendency and may be defined as the value, which we get by dividing the total of the values of various given items in a series by the total number of items. We can work it out as follows

$$\text{Mean, } \bar{x} = \frac{\sum xij}{N}$$

Mean is the simplest measurement of central tendency and is a widely used measure. Its chief use consists in summarizing the essential features of a series and in enabling data to be compared. It is a relatively stable measure of central tendency, but it suffers from some limitations of unduly affected by extremes; it may not coincide with actual value of an item in a series, and it may lead to strong impressions, particularly when the item values are not given with the average. However, mean is better than other average, especially in economic and social studies where direct quantitative measurements are possible.

Standard Deviation

Standard deviation is the most widely used measure of dispersion of a series and is commonly denoted by the symbol σ(pronounced as sigma). Standard deviation is the square root of the arithmetic mean of the square of deviations, the deviations being measured from the arithmetic mean of distribution. It is less affected by sampling errors and is more stable measure of dispersion. It is worked out as follows,

$$\text{Standard deviation}, \sigma = \frac{\sqrt{\sum(x-x)}}{N-1}$$

Coefficient of Variation

A measure of variation which is independent of the unit of measurement is provided by coefficient of variation. Being unit free, this is useful for computation of variability between different populations. The coefficient of variation is standard deviation expressed as percentage of the mean and is measured by the formula,

Coefficient of correlation

When an increase or decrease in one variable is accompanied by an increase or decrease in other variable, the two are said to be correlated and the phenomenon is known as correlation. Correlation coefficient (r) is a measure of the relationship between two variables, which are at the interval or ratio level of measurement and are linearly related. A Spearman's coefficient of correlate

The value of 'r' lies between +1 to -1. Positive values of r indicate that positive correlation between the two variables (i.e. change in both variables takes place in the same direction), whereas negative values of 'r' indicate negative correlation i.e. changes in the two variables taking place in opposite direction. A zero value of 'r' indicates that there is no association between the two variables. When 'r' (+) 1, it indicates perfect positive correlation and when it is (-) 1, it indicates perfect negative correlation, meaning thereby that variations in independent variable (x) explain 100 per cent of the variations in the dependent variable (y). We can also say that for a unit change in independent variable, if there happens to be constant change in the dependent variable in the same direction, the correlation will be termed as perfect positive. But if such change occurs in the opposite direction, the correlation will be termed as perfect negative. The values of 'r' nearer to =1 or -1 indicates high degree of correlation between the two variables.

Stepwise Multiple Regression

Stepwise regression is a variation of multiple regressions which provides a means of choosing independent variables that in the best prediction possible with the fewest independent variables. It permits the user to solve a sequence of one or more multiple linear regression problems by stepwise application of the least square method. At each step in the analysis, a variable is added or removed which results in the greatest production of error sum of square (Burroghs corporation, 1975).

According to Drapper and Smith (1981), the method of stepwise multiple regression analysis is to insert variables in turn until the regression equation is satisfactory. The order of insertion is determined by using the partial correlation coefficient as the measure of the importance of variables not yet in the equation.

The program, according to Burrogh's Corporation (1975), first forms a correlation matrix, finds the best predictor (the independent variable having the highest correlation with criterion variable) and performs a regression analysis with this predictor. Then, the second best predictor (independent), and so on. At any given stage, the group of predictors being used is not necessarily the best group of that size (i.e. the particular group of independent variables does not necessarily have the highest multiple correlation with the criterion that any group of this size does). Rather, this group contains the variables that have the highest individual correlation with the criterion.

Stepwise regression is the step-by-step iterative construction of a model that involves the selection of independent variables to be used in a final model. It involves adding or removing potential explanatory variables in succession and testing for statistical significance after each iteration. Stepwise regression is a method that iteratively examines the statistical significance of each independent variable in a linear regression model.

The underlying goal of stepwise regression is, through a series of tests (e.g. F-tests, to find a set of independent variables that significantly influence the dependent variable. Stepwise regression can be achieved either by trying out one independent variable at a time and including it in the regression model if it is or by including all potential independent variables in the model and eliminating those that are not statistically significant. There are three approaches to stepwise regression:

- **Forward selection-** It begins with no variables in the model, tests each variable as it is added to the model, then keeps those that are deemed most statistically significant—repeating the process until the results are optimal.
- **Backward elimination-** This starts with a set of independent variables, deleting one at a time, then testing to see if the removed variable is statistically significant.
- **Bidirectional elimination-** This is a combination of the first two methods that test which variables should be included or excluded.

Path Analysis

The term 'path analysis' was first introduced by the biologist, Sewall Wright in 1934 in connection with decomposing the total correlation between any two variables in a causal system. The technique of path analysis is based on a series of multiple regression analysis with the added assumption of causal relationship between independent and dependent variables. Path analysis makes use of standardized partial regression coefficient (known as beta weights) as effect coefficients. In linear additive effects are assumed, and then through path analysis a simple set of equations can be built up showing how each variable depends on preceding variables. The main principle of path analysis is that any correlation coefficient between two variables, or a gross or overall measure of empirical relationship can be decomposed into a series of paths: separate paths of influence leading through chronologically intermediate variable to which both the correlated variables have linked.

The merit of path analysis in comparison to correlation analysis is that it makes possible the assessment of the relative influence of each antecedent or explanatory variable on the consequent or criterion variables by first making explicit the assumptions underlying the causal connections and then by elucidating the indirect effect of the explanatory variables.

Canonical Covariate Analysis

A canonical correlation is the correlation of two canonical variables, one representing a set of independent variables, the other a set of dependent variables. Each set may be considered a latent variable based on measured indicator variables in its set. The canonical correlation is optimized such that the linear correlation between the two latent variables is maximized. Canonical correlation is used for many to many relationships. There may be more than one such linear correlation relating the two set of variables, with each such correlation representing a different dimension by which the independent set of variables is related to the dependent set. The purpose of canonical correlation is to explain the relation of the two sets of variables not to model the individual variables. In addition to asking how strong the relationship is between two latent variables, canonical correlation is useful in determining how many dimensions are needed to account for that relationship. Canonical correlation finds the linear combination of variables that produces the largest correlation with second set of variables.

Canonical correlation is a member of multiple general linear hypothesis (MLGH) family and shares many of assumptions of multiple regression such as linearity of relationship, homoscedasticity (same level of relationship for the full range of data), interval or near interval data, untruncated variables, proper specification of model, lack of high multi-collinearity and multivariate normality for purpose of hypothesis testing.

Factor Analysis

The term factor analysis refers to a set of analytical techniques designed to reduce data into smaller, meaningful groups based upon their inter-correlations or shared variance. The assumption is that those items or variables that are correlated must be measuring a similar factor or trait or construct. In the case where only a few variables are used, the researcher may be able to determine groupings by simply observing the content of each variable; however, for large data sets and/or more ambiguous items, this task would be formidable. Factor Analysis reduces the information in a model by reducing the dimensions of the observations. This procedure has multiple purposes. It can be used to simplify the data, for example reducing the number of variables in predictive regression models. If factor analysis is used for these purposes, most often factors are rotated after extraction. Factor analysis has several different rotation methods—some of them ensure that the factors are orthogonal. Then the correlation coefficient between two factors is zero, which eliminates problems of multicollinearity in regression analysis.

Factor analysis is also used in theory testing to verify scale construction and operationalizations. In such a case, the scale is specified upfront and we know that a certain subset of the scale represents an independent dimension within this scale. This form of factor analysis is most often used in structural equation modeling and is referred to as Confirmatory Factor Analysis. For example, we know that the questions pertaining to the big five personality traits cover all five dimensions N, A, O, and I. If we want to build a regression model that predicts the influence of the personality dimensions on an outcome variable, for example anxiety in public places, we would start to model a confirmatory factor analysis of the twenty questionnaire items that load onto five factors and then regress onto an outcome variable.

Factor analysis can also be used to construct indices. The most common way to construct an index is to simply sum up the items in an index. In some contexts, however, some variables might have a greater explanatory power than others. Also sometimes similar questions correlate so much that we can justify dropping one of the questions completely to shorten questionnaires. In such a case, we can use factor analysis to identify the weight each variable should have in the index.

Factor analysis is a data reduction technique that can reduce the number of items by grouping them and by examining the content of the items in each group one can determine the structure or composition of each group thereby giving a better explanation of the data. It is important to note that factor analysis is not used in prediction or explaining the relationship between different sets of variables, nor is it used to determine group differences. The goal is to explain the underlying structure or composition of the data; therefore we are dealing only with one set of variables.

Two types of factor analysis exist. The first, *exploratory factor analysis* (EFA) is used to explore or derive the underlying factor structure of a data matrix often without regard to theory. The purpose of EFA is to determine if underlying factors exist within a data set, and if so, what those factors are. The researcher does not necessarily need to have any expectations or theory beforehand—it can simply be "exploratory." Two types of EFA are commonly used. *Principle components analysis* (PCA) tries to account for all variance among the variables/items and so includes both shared variance and unique/error variance of each variable/item. *Principle factor analysis* (PFA) accounts for only shared variance of each variable/item. Currently, most journals prefer PFA.

The second, *confirmatory factor analysis* (CFA), is used to test a priori theory. The researcher specifies what factors exist and what variables/items constitute each factor and then "orders" these parameters into a data set to determine if indeed the factors and variables/items describe or fit the data. CFA is most commonly conducted through Structural Equation Modelling (SEM). The following discussion is on factor analysis pertaining to EFA.

The main goal of factor analysis is to explain as much variance as possible in a data set by using the smallest number of factors (groupings of variables based on high inter-correlations) and the smallest amount of items or variables within each factor. Inherently one balances explained variance with simplicity. Critical to this technique is that one wants to ensure that the variance left out of the solution is primarily error variance. In EFA, the first factor derived is a linear composite of all variables/items such that it maximizes the amount of total variance explained (or extracted)—no other linear combination will extract as much variance. The proportion of variance extracted is called the *eigen value.* A second factor which is orthogonal to the first (uncorrelated) is then derived from the remaining variance, and its eigen value will be derived. Like the first factor, the second factor will be a linear composite of all variables/items. This process continues until all variance has been extracted, and the number of extractions is equal to the number of original variables/items.

Interpretation: Interpretation involves three steps:

- The researcher determines the number of factors to keep. Several decision rules can be employed but generally keep those factors with eigen values greater than 1.0. The reason is that any given variable has a variance equal to 1.0 (since variables are standardized, the std. dev. = 1.0, and variance is the std. dev. Squared) which means that eigen values should explain more variance than at least one variable/item. In a satisfactory EFA, the total variance of the retained eigen values should be greater than 70%.
- Once the number of factors has been determined, the researcher then determines which variables/items "load" on each factor. This is determined by the coefficient of the variable/item. Because each factor is a linear composite of all variables/items, each variable/item will have a different coefficient for each factor. This coefficient is actually a structure coefficient since it is the correlation between the variable/item and the factor; however, it is called the factor loading or the factor structure coefficient. The higher the coefficient, the greater the variable/item's contribution to the factor, and the square of the coefficient is the amount of variance of the factor that is explained by the variable/item. In general, the desired outcome is that each variable/item will have a large loading on only one factor and small loadings on the remaining factors.

Before actually determining the composition of each factor, the axes which represent factor dimensions can be rotated geometrically so that the new set of axes are positioned closer to their respective factor variables/items. These axes are similar to those in a Cartesian coordinate system. For example, in a two-factor model, every variable/item will have a factor loading value for Factor I and for Factor II. Most will have a relatively high loading value (>.5) on one factor and relatively small loading on the other factor (< .5). Each factor is an axis in space and the factor loading of a variable/item is the coordinate value on that axis. If variable X1 has a factor loading of .63 on FI, and .22 on FII, then its coordinate point is (.63, .22). Also, the distance of X1 from the origin can be calculated by the Pythagorean theorem (add .63 squared and .22 squared and take the square root). By rotating the axes, one can increase the value along the FI dimension and decrease the FII dimension value as long as the distance of X1 to the origin remains unchanged. This strengthens the association of X1 with FI while decreasing its association to FII. This makes interpretation of the factor structure (composition) simpler. Usually when rotating the axes, the researcher assumes that the factors are truly orthogonal and so an orthogonal rotation is used, that is the axes remain at 90 degrees to one another. This is easy to visualize in a two-factor model. If F1 is rotated 35 degrees clockwise, FII must also be rotated 35 degrees clockwise. However, sometimes the factors are somewhat correlated and so the rotation that occurs is based upon the correlation between the factors. The cosine of the angle between the axes is equal to the correlation. By the way, this rule holds true when orthogonal rotation is used because the angles are at 90 degrees and the cosine of 90 degrees is 0.

- 3) With the output arranged in a table where the columns are the factors and the rows are the variables/items, one then highlights or underscores the loadings under each factor that are greater than .50. Once all contributing variables on all factors have been identified, then one must determine the content of each factor and assign an appropriate name for each. This is done by analyzing the content or general theme of the variable/items that are highlighted.

Principal Component Analysis

Principal component analysis (PCA) involves a mathematical procedure that transforms a number of (possibly) correlated variables into a (smaller) number of uncorrelated variables called *principal components.* The first principal component accounts for as much of the variability in the data as possible, and each succeeding component accounts for as much of the remaining variability as possible.

Objectives of principal component analysis:

- To discover or to reduce the dimensionality of the data set.
- To identify new meaningful underlying variables.

Traditionally, principal component analysis is performed on the matrix or on the matrix. These matrices can be calculated from the data matrix. The covariance matrix contains scaled. A correlation matrix is like a covariance matrix but first the variables, i.e. the columns, have been standardized. We will have to standardize the data first if the variances of variables differ much, or if the units of measurement of the variables differ. The mathematical technique used in PCA is called eigen analysis: we solve for the eigen values and eigen vectors of a square symmetric matrix with sums of squares and cross products. The eigenvector associated with the largest eigen value has the same direction as the first principal component. The eigenvector associated with the second largest eigen value determines the direction of the second principal component. The sum of the eigen values equals the trace of the square matrix and the maximum number of eigenvectors equals the number of rows (or columns) of this matrix.

There are two methods to help you to choose the number of components to keep. Both methods are based on relations between the eigen values.

- Plot the eigen values, If the points on the graph tend to level out (show an "elbow"), these eigen values are usually close enough to zero that they can be ignored.
- Limit the number of components to that number that accounts for a certain fraction of the total variance. (Acharya et al, 2020)

Cluster Analysis

The purpose of cluster analysis is to reduce a large data set to meaningful subgroups of individuals or objects. The division is accomplished on the basis of similarity of the objects across a set of specified characteristics. Outliers are a problem with this technique, often caused by too many irrelevant variables. The sample should be representative of the population, and it is desirable to have uncorrelated factors. There are three main clustering methods hierarchical, this is a tree like process appropriate for smaller data set; non-hierarchical, which requires specification of the number of clusters apriori; and a combination of both. There are four main rules for developing clusters – the clusters should be different, they should be reachable, they should be measurable and the clusters should be profitable (big enough to matter). This is a great tool for market segmentation.

Artificial Neural Network Analysis

ANNs are computing systems inspired by the biological neural network that constitute animal brains. Such systems learn (progressively improve performance on) tasks by considering examples, generally without task specific programming. An ANN is based on a collection of connected units or nodes called artificial neurons (analogous to biological neurons in an animal brain). Each connection (analogous to a synapse) between artificial neurons can transmit a signal from one to another. The artificial neuron that receives the signal can process it and then signal artificial neurons connected to it. In common ANN implementation, the signal at a connection between artificial neurons is a real number, and the output of each artificial neuron is calculated by a non-linear function of the sum of its inputs. Artificial neurons and connections typically have a weight that adjusts as learning proceeds. The weight increases or decreases the strength of the signal at a connection. Artificial neurons may have a threshold such that only if the aggregate signal crosses that threshold is the signal sent. Typically, artificial neurons are organized in layers. Different layers may perform different kind of transformations on their inputs. Signals travel from the first (input), to the last (output) layer, possibly after traversing the layers multiple times. The original goal of the ANN approach was to solve problems in the same way that a human brain would. Over time, attention focused on matching specific mental abilities, leading to deviations from biology. ANNs have been usedon a variety of tasks, including computer vision, speech recognition, machine translation, social network filtering, playing board and video games and medical diagnosis.

Chapter-6

RESEARCH AND CASE STUDIES

It helps visualizing the distributive characters of performing variables in terms of their deviation, variance and consistency.

Table:1 presents the descriptive statistics of independent variables for the clear visualization of the distribution, departure and central tendencies variables.

The distribution of 'age' (x_1) depicts that the avg. age of the respondents stands at 50.76 years or around 51 years. So, most of the respondents are above mid-age category with SD=8.52 and CV=16.79 percent. This indicates that the distribution of the variables is more consistent.

The distribution of 'education'(x_2) that the avg. education of the respondents stands at 11.11 or around 12 with SD=2.01 and CV=18.13%.Maximum is 15 and minimum is 7.This implies that the distribution of the variables more consistent.

The distribution of 'family members' (x_3) indicates that the avg. family members of the respondents stands at 6.10 or around 7 with SD=1.43 and CV=23.48%.The maximum is 10 and the minimum is 4.This implies that the distribution of the variables more consistent.

The distribution of 'training and capacity building'(x_4) implies that the avg. Training and capacity building of the respondents stands at 2.99 or around 3 with SD=1.33 and CV=44.59%.Maximum is 8 and minimum is 1.This indicates the distribution of the variables medium consistent.

The distribution of 'Family education score' (x_5) indicates that the avg. family education score of the respondents stands at 8.59 or around 9 with SD=1.74 and CV=20.29%.Maximum is 13 and minimum is 3.5.This implies that the distribution of the variables is more consistent.

Table1. Descriptive Statistics of Independent Variable with respect to Minimum, Maximum, Range, Mean, Standard Deviation of Values, Variance, and Coefficient of Variance:

Variable	*Minimum*	*Maximum*	*Mean*	*Std. Deviation*	*Variance*	*CV (%)*
Age (x_1)	28.00	70.00	50.76	8.52	72.64	16.79
Education (x_2)	7.00	15.00	11.11	2.01	4.06	18.13
Family members (x_3)	4.00	10.00	6.10	1.43	2.05	23.48
Training and capacity building(x_4)	1.00	8.00	2.99	1.33	1.77	44.59
Family education score (x_5)	3.50	13.00	8.59	1.74	3.04	20.29
Holding size (x_6)	2.00	15.00	5.55	2.30	5.29	41.40
Homestead land (x_7)	0.50	20.00	6.83	2.86	8.19	41.85
No. of fragments (x_8)	1.00	5.00	2.59	0.81	0.66	31.53
Cultivable land area (x_9)	1.50	15.00	4.52	1.84	3.38	40.67
Cropping intensity (x_{10})	25.00	100.00	73.42	21.61	467.08	29.43
Annual Income (x_{11})	21580.00	3045000.00	325795.42	496923.63	246933102867.36	152.52
Yield (x_{12})	107.00	12000.00	3494.09	3317.47	11005641.93	94.94
Social participation (X_{13})	1.00	22.00	5.74	3.15	9.92	54.85
Constraint perception (x_{14})	1.00	6.00	2.52	1.20	1.46	47.93
Suggestions for improvement (x_{15})	2.00	6.75	3.76	0.93	0.88	24.92
Communication through mobile (x_{16})	1.00	5.00	2.46	0.88	0.79	36.02
No. of fragments (x_{17})	1.50	15.00	5.17	2.14	4.60	41.43

Variable	Minimum	Maximum	Mean	Std. Deviation	Variance	CV (%)
No. of stakeholder in different practices (x_{18})	28.00	868.00	151.98	136.31	18580.42	89.68
Total Production (x_{19})	1650.00	940000.00	36571.88	137735.11	18970962274.52	376.61
Family expenditure (x_{20})	100000.00	4454000.00	627995.63	597077.77	356501871882.11	95.07
Energy consumption (x_{21})	17800.00	30200.00	26628.57	2397.62	5748598.90	9.00
Nutrition and Health Status (x_{22})	6500.0000	30000.0000	16461.904762	5100.7612734	26017765.568	30.98524
Average wage/Day (y_1)	244.0000	135000.0000	2824.123564	13442.3607909	180697063.634	475.9835
Family farm income (y_2)	27000.0000	1518000.0000	338406.095238	262506.5375871	68909682275.952	77.57146
Productivity (y_3)	107.0000	35000.0000	3615.307714	4532.3563829	20542254.382	125.3657
Income per bigha(y_4)	21580.0000	3045000.0000	329607.142857	504109.8437969	254126734612.912	152.9426
Man days (y_5)	28.0000	868.0000	145.104762	133.1809010	17737.152	91.78258
Ecology perception (y_6)	2.1000	9.2000	4.920571	1.2826998	1.645	26.06811

The distribution of 'Holding size'(x_6) indicates that the avg. holding size of the respondents stands at 5.55 with SD=2.3 and CV=41.40%.This implies that the distribution of the variables medium consistent.

The distribution of the 'Homestead'(x_7) indicates that the avg. homestead of the respondents stands at 6.83 with SD=2.86 or around 3 and CV=41.85%.Maximum is 20 and minimum is 0.5.This implies that the distribution of the variables medium consistent.

The distribution of 'no. of fragments'(x_8) indicates that the avg. no. of fragments is 2.59 or around 3 with SD=0.816 or around 1 and CV=31.53%.Maximum is 5 and minimum is 1.This implies that the distribution of the variables medium consistent.

The distribution of 'Cultivable land area'(x_9) indicates that the avg. no. of cultivable land area is 4.52 or around 5 with SD=1.84 and CV=40.67%.Maximum is 15 and minimum is 1.5.This implies that the distribution of the variables medium consistent.

The distribution of 'cropping intensity' (x_{10}) indicates that the avg. no. of cropping intensity is 73.42 with SD=21.61 and CV=29.43%.Maximum is 100 and minimum is 25.This implies that the distribution of the variables more consistent.

The distribution of the 'annual income'(x_{11}) indicates that the avg. no. of annual income is 325795.42 or around 33 with SD=496923.63 and CV=152.52%.Maximum is 100 and minimum is 25.This implies that the distribution of the variables less consistent

The distribution of 'yield' (x_{12}) indicates that the avg. no. of distribution of yield is 3494.09 or around 35 with SD=3317.47 or around 34 and CV=94.94%. Macimum is 12000 and minimum is 107.This implies that the distribution of the variables less consistent.

The distribution of 'social participation' (x_{13}) indicates the avg. no. of distribution of social participation is 5.74 or around 6 with SD=3.15 and CV=54.85%.This implies that the distribution of the variables medium consistent.

The distribution of 'constraint perception' (x_{14}) indicates that mean value of constraint perception is 2.52 or around 3 with standard deviation (SD) = 1.2 and CV=47.93%. Maximum is 6 and minimum is 1.This implies that the distribution of variables is of medium consistent.

The constraint to agricultural management stands in between the high and low planks.

The distribution of 'suggestions for improvement'(x_{15}) indicates that the avg. no. of suggestions of improvement is 3.76 with SD=0.93 and CV= 24.92%. Maximum is 0.75 and minimum is 2. This implies that the avg. no. of variables is more consistent.

The distribution of 'communication through mobile' (x_{16}) indicates that the avg. no. of distribution of communication through mobile is 2.46 with SD=0.88

or around 1 and CV=36.02.Maximum is 5 and minimum is 1.This implies that the distribution of the variables are medium consistent.

The distribution of 'No. of Fragments'(x_{17}) indicates that the avg. no. of fragments is 5.17 with SD=2.14 and CV=41.43%.Maximum is 15 and minimum is 1.5.This implies that the distribution of no. of fragments is medium consistent.

This indicates that the no of fragments per holding is very high on an average. And, it will make the farm management energy and resource prodigal.

The distribution of 'No. of Stakeholders involved in different practices' (x_{18}) indicates that the avg. no. of stakeholders in different practices is 151.98 or around 152 with SD=136.31 and CV=89.68%.Maximum is 868 and minimum is 28.This implies that the distribution of the variables is less consistent.

The distribution of 'total production' (x_{19}) indicates that the avg. no. of total production is 36571.88 with SD=1357735.11 and CV=376.61.Maximum is 94 and minimum is 1650.This implies that the avg. no. of variables less consistent.

The distribution of 'Family expenditure'(x_{20}) indicates that the avg. no. of family expenditure is 627995.63 with SD=597077.77 and CV=95.07.Maximum is 4454000 and minimum is 100000.This implies that the avg. no. of variables is less consistent.

The distribution of 'energy consumption' (x_{21}) indicates that the avg. no. of energy consumption is 26628.57 with SD=2397.62 and CV=9%.This implies that the avg. no. of variables is more consistent.

Table 2: Coefficient of Correlation (r): Average wage/day (y_1) vs. 22Independent Variables (x_1-x_{22})

Sl. No.	Independent Variables	'r' Value	Remarks
1	Age (x_1)	0.09	
2	Education (x_2)	0.02	
3	Family members (x_3)	0.06	
4	Training and capacity building (x_4)	-0.11	
5	Family education score (x_5)	-0.02	
6	Holding size (x_6)	0.17	
7	Homestead land (x_7)	0.07	
8	No. of fragments (x_8)	0.03	
9	Cultivable land area (x_9)	0.03	
10	Cropping intensity (x_{10})	-.23	*
11	Annual Income (x_{11})	0.08	
12	Yield (x_{12})	-0.10	
13	Social participation (x_{13})	-0.12	
14	Constraint perception (x_{14})	.19	*
15	Suggestions for improvement (x_{15})	-0.12	

Sl. No.	Independent Variables	'r' Value	Remarks
16	Communication through mobile (x_{16})	0.03	
17	No. of fragments (x_{17})	0.02	
18	No. of stakeholder in different practices (x_{18})	0.06	
19	Total Production (x_{19})	-0.04	
20	Family expenditure (x_{20})	-0.03	
21	Energy consumption (x_{21})	-0.15	
22	Nutrition and Health Status (x_{22})	0.03	

**Correlation is significant at the 0.01 level

*Correlation is significant at the 0.05 level

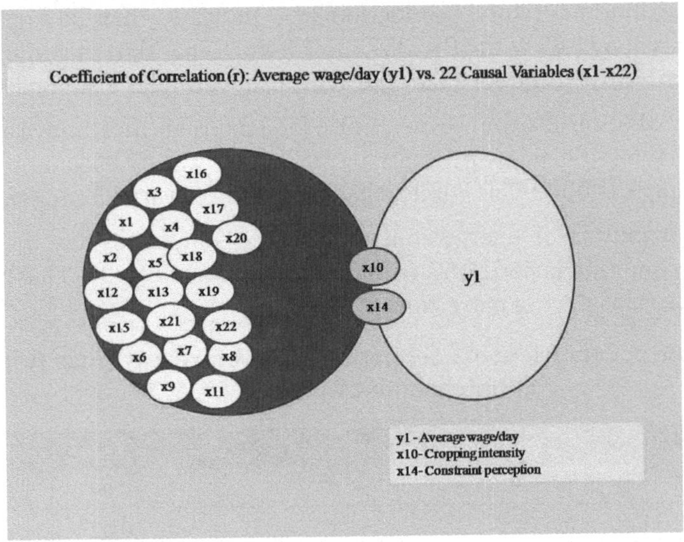

Model 1

Table-2presents the coefficient of correlation between wages day (y_1) and 22 exogenous variables selected for the study:

It is interesting to note that cropping intensity has recorded a significant net a negative correlation with wage and employment.This indicates that escalation of cropping intensity does not reflect on a positive change in wage and employment.

Due to brunt of pandemic and prevailing fear psychosis,the human labours have been found in less number on the field than the usual time. This has affected both productivity for the owner of the land and wages for the agricultural labourers.The site of story is that in this area of alluvial zone (Nadia and Hooghly), agricultural intensification has been associated with mechanization. The intensity of farm mechanization has been ironically linked with the drifting of agricultural labours off the farm.

Constraint perception (x_{14}) has recorded a positive influence on wage and employment(y_1).It signifies that to a better understanding on the constraints confronted by him,he perhaps able to solve it and can successfully manage his farm with higher wage generation and employability.

Table:3: Multiple Regression Analysis: Average wage/day (y_1) Vs. 22 Causal Variables (x_1-x_{22})

Sl. No.	*Variables*	*Reg. Coeff. B*	*S.E. B*	*Beta*	*t Value*
1	Age (x_1)	0.06	0.05	0.12	1.05
2	Education (x_2)	0.02	0.05	0.05	0.42
3	Family members (x_3)	0.03	0.08	0.08	0.45
4	Training and capacity building (x_4)	-0.02	0.06	-0.04	-0.44
5	Family education score (x_5)	0.01	0.07	0.03	0.25
6	Holding size (x_6)	0.24	0.11	0.43	2.16
7	Homestead land (x_7)	0.00	0.09	0.00	0.01
8	No. of fragments (x_8)	-0.02	0.05	-0.04	-0.44
9	Cultivable land area (x_9)	-0.22	0.11	-0.30	-1.95
10	Cropping intensity (x_{10})	-0.10	0.04	-0.30	-2.62
11	Annual Income (x_{11})	0.20	0.13	0.32	1.44
12	Yield (x_{12})	-0.01	0.04	-0.05	-0.45
13	Social participation (x_{13})	-0.11	0.07	-0.16	-1.54
14	Constraint perception (x_{14})	0.11	0.04	0.27	2.49
15	Suggestions for improvement (x_{15})	-0.03	0.05	-0.07	-0.67
16	Communication through mobile (x_{16})	0.08	0.05	0.18	1.65
17	No. of fragments (x_{17})	0.00	0.08	0.00	0.02
18	No. of stakeholder in different practices (x_{18})	-0.31	0.14	-0.49	-2.11
19	Total Production (x_{19})	-0.07	0.08	-0.11	-0.99
20	Family expenditure (x_{20})	-0.01	0.08	-0.02	-0.18
21	Energy consumption (x_{21})	-0.06	0.06	-0.12	-1.04
22	Nutrition and Health Status (x_{22})	-0.06	0.08	-0.14	-0.76

R square: 27.10%

The standard error of the estimate: 0.9080850

Table:3 presents the multiple regression analysis (full model) between wage and employment (y1) versus 22 causal variables.

The table shows that all the 22 variables conglomerated together have contributed 47% of variance in the consequent variable, wages and employment.

Now, in order to identify the dominant variables the data set has undergone the stepwise regression analysis as presented in table 3.

The stepwise regression analysis depicts that the variable cropping intensity and constraint perception have been retained at the last step.

So, improvement in wage and employment are hovering around two marker variables i.e. cropping intensity and constraint perception.

So, we need to concentrate on not only to increase the cropping intensity, but also to improve the scope and opportunity of wage and employment generation; on the other hand, the plethora of constraints like poor market communication, poor supply chain functions, lack of soft skill and different institutional barriers making credit input and knowledge hardly accessible need to be removed. Else, question of hungry and poor rural people may be found by the site of green crop field.

Table:4: Stepwise Regression Analysis: Average wage/day (y_1) Vs. 22 Causal Variables (x_1-x_{22})

Sl. No.	*Variables*	*Reg. coeff. B*	*S.E. B*	*Beta*	*t value*
1	Cropping intensity (x_{10})	-0.09	0.03	-0.25	-2.72
2	Constraint perception (x_{14})	0.09	0.04	0.22	2.36

R square: 10.50%

The standard error of the estimate:0.974703

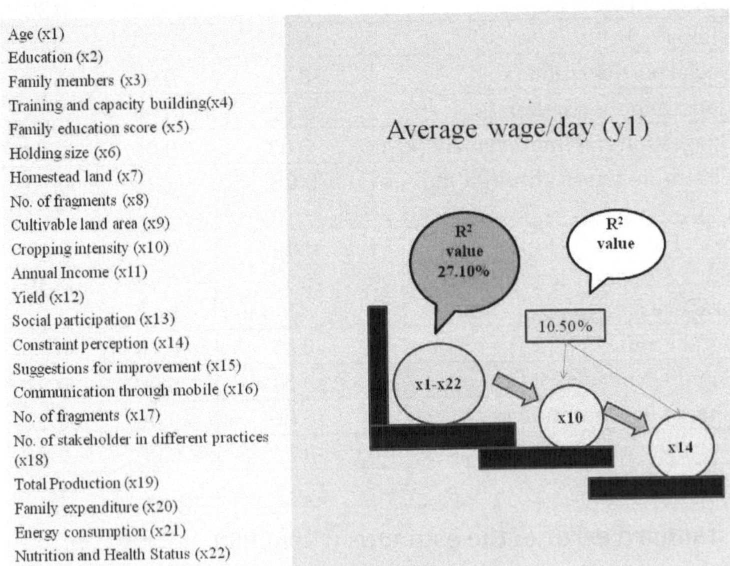

Model 2

Table:4 presents the stepwise regression analysis where two variables: cropping intensity and constraint perception have been retained at the last step With the increase in cropping intensity, the number of crop enterprises as well as farm operations has increased. So, it has substantially contributed to wage.

Similarly, when there has been a better perception on constraints, its management will be easier. With the better redress of constraints, the productivity and wage both would go up.

Table:5: Path Analysis: Decomposition of Total Effect into Direct, Indirect and Residual Effect: Average wage/day (y_1) vs. 22 exogenous variables (x_1-x_{22})

Sl. No.	Variables	Total Effect	Direct Effect	Indirect Effect	Highest Indirect Effect
1	Age (x_1)	0.09	0.12	-0.02	-0.14 (x_{18})
2	Education (x_2)	0.02	0.05	-0.02	0.08 (x_6)
3	Family members (x_3)	0.06	0.08	-0.02	0.23 (x_6)
4	Training and capacity building (x_4)	-0.11	-0.05	-0.06	0.09 (x_{18})
5	Family education score (x_5)	-0.02	0.03	-0.05	-0.08 (x_9)
6	Holding size (x_6)	0.17	0.43	-0.25	-0.27 (x_{18})
7	Homestead land (x_7)	0.07	0.00	0.06	0.19 (x_6)
8	No. of fragments (x_8)	0.03	-0.05	0.08	0.17 (x_6)
9	Cultivable land area (x_9)	0.03	-0.30	0.34	0.24 (x_6)
10	Cropping intensity (x_{10})	-0.23	-0.30	0.07	0.09 (x_9)
11	Annual Income (x_{11})	0.08	0.32	-0.23	-0.43 (x_{18})
12	Yield (x_{12})	-0.10	-0.05	-0.05	0.10 (x_{18})
13	Social participation (x_{13})	-0.12	-0.16	0.04	0.04(x_{10})
14	Constraint perception (x_{14})	0.19	0.27	-0.07	-0.08 (x_{18})
15	Suggestions for improvement (x_{15})	-0.12	-0.07	-0.04	0.05 (x_{10})
16	Communication through mobile (x_{16})	0.03	0.18	-0.14	0.15 (x_{18})
17	No. of fragments (x_{17})	0.02	0.00	0.01	0.25 (x_6)
18	No. of stakeholder in different practices (x_{18})	0.06	-0.49	0.56	0.28 (x_{11})
19	Total Production (x_{19})	-0.04	-0.11	0.06	0.08 (x_{18})

Sl. No.	Variables	Total Effect	Direct Effect	Indirect Effect	Highest Indirect Effect
20	Family expenditure(x_{20})	-0.03	-0.02	-0.01	-0.07 (x_{18})
21	Energy consumption (x_{21})	-0.15	-0.12	-0.03	0.12 (x_{18})
22	Nutrition and Health Status (x_{22})	0.03	-0.14	0.17	0.20 (x_6)

Residual effect:0.72854

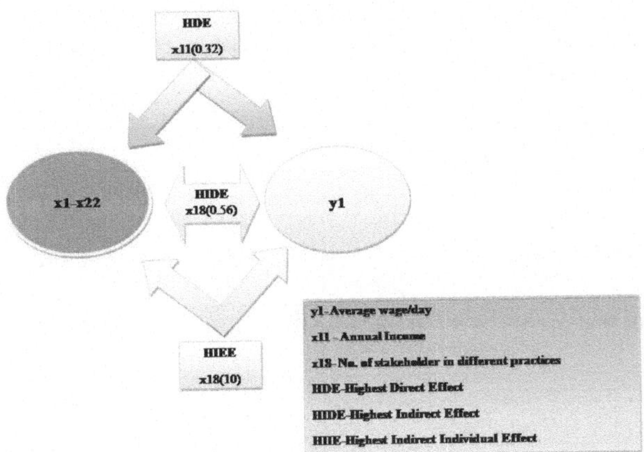

Model

Table:5 presents the path analysis by decomposing the total effect (r value) into direct,indirect and residual effect.

It has been found that income is the most important determinant for a farmer from his farm as to whether his income will go after generations the wage and employment with higher income he will be able to hire more labours and can ensure better input and management.

The highest indirect effect has been exerted by the variable no. of stakeholders involved in different practices. This implies that there has been an orchestration of multi-stakeholder participation making farm economics successfully generating higher income with a simultaneous creation of wage and employment.

Table:6: Coefficient of Correlation (r): (y_2) Vs. 22Independent Variables

Sl. No.	Independent Variables	'r' Value	Remarks
1	Age (x_1)	0.149	
2	Education (x_2)	0.152	
3	Family members (x_3)	0.125	

Sl. No.	Independent Variables	'r' Value	Remarks
4	Training and capacity building (x_4)	-0.029	
5	Family education score (x_5)	0.115	
6	Holding size (x_6)	.438**	**
7	Homestead land (x_7)	.256**	**
8	No. of fragments (x_8)	0.134	
9	Cultivable land area (x_9)	.205*	*
10	Cropping intensity (x_{10})	-0.002	
11	Annual Income (x_{11})	.268**	**
12	Yield (x_{12})	0.016	
13	Social participation (x_{13})	0.055	
14	Constraint perception (x_{14})	0.015	
15	Suggestions for improvement (x_{15})	0.153	
16	Communication through mobile (x_{16})	-.208*	*
17	No. of fragments (x_{17})	.250**	**
18	No. of stakeholder in different practices (x_{18})	.272**	**
19	Total Production (x_{19})	0.128	
20	Family expenditure (x_{20})	-0.130	
21	Energy consumption (x_{21})	-.271**	**
22	Nutrition and Health Status (x_{22})	0.180	

**Correlation is significant at the 0.01 level

*Correlation is significant at the 0.05 level

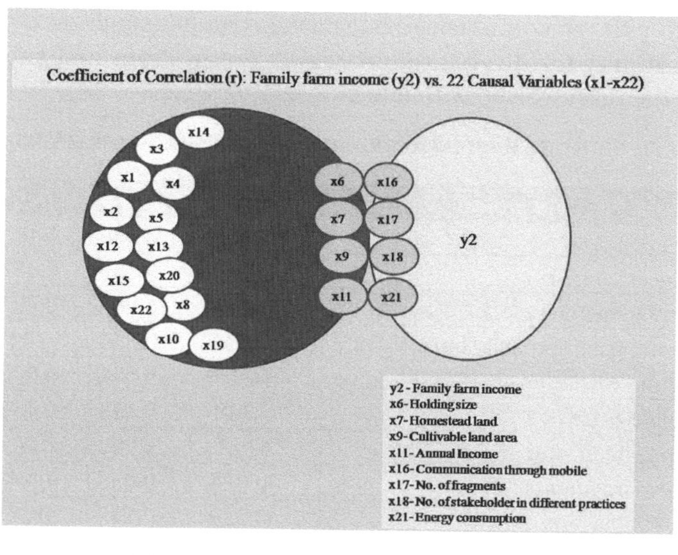

Model 3

Table:6presents the coefficient of correlation between Family Farm Income and 22 exogenous variables

Holding size(x_6) has been found to exert significant and positive influence on family farm income. It is quite obvious that the bigger is the holding size, the higher has been the Family Farm Income.

Homestead land cultivable area has also exerted positive and significant influence in augmenting income from Family Farm.

So, it has been obvious that land resources are still the strongest determinant in ushering Family Farm Income.

In an enterprising agriculture, ability to invest is one of the determinants which can make Farm management and production proficient and profitable. Other source of income of a Family Farm extends immense support to enhance investment ability, risk bearing capacity and thus helps farming a better Farm income.

Communication through mobile (x_{16}) and Energy consumption (x_{21}) however have recorded negative and significant influence on Family Farm Income while ecological level of energy consumption has contributed positive change in Family Farm Income, Mobile telephonic could not record any impressive effect on improving Farm Income.

On the other hand, no. of fragments (x_{17}) and no. of stakeholder in different practices (x_{18}) have positively influenced. To improve the Family Farm Income(x_2) small fragment of land here have been found to have higher proficiency in generalizing Family Farm Income and the more number of stakeholder involved in farm production process have also exerted positive influence on Family Farm Income.

More no. of fragments of land being managed under single ownership, offers better opportunity for diversification with different crops and enterprises by minimising the risk involved with monolithic enterprise.

Table:7: Multiple Regression Analysis: (y_2) Vs. 22 Causal Variables

Sl. No.	*Variables*	*Reg. Coeff. B*	*S.E. B*	*Beta*	*t Value*
1	Age (x_1)	0.061	0.058	0.121	1.055
2	Education (x_2)	0.021	0.050	0.052	0.426
3	Family members (x_3)	0.037	0.082	0.087	0.450
4	Training and capacity building (x_4)	-0.026	0.060	-0.049	-0.443
5	Family education score (x_5)	0.018	0.071	0.032	0.253
6	Holding size (x_6)	0.248	0.115	0.430	2.160
7	Homestead land (x_7)	0.001	0.092	0.002	0.012
8	No. of fragments (x_8)	-0.024	0.055	-0.049	-0.441
9	Cultivable land area (x_9)	-0.228	0.117	-0.305	-1.951
10	Cropping intensity (x_{10})	-0.108	0.041	-0.306	-2.624
11	Annual Income (x_{11})	0.200	0.139	0.322	1.441

Sl. No.	Variables	Reg. Coeff. B	S.E. B	Beta	t Value
12	Yield (x_{12})	-0.018	0.041	-0.050	-0.451
13	Social participation (x_{13})	-0.112	0.073	-0.165	-1.548
14	Constraint perception (x_{14})	0.117	0.047	0.277	2.496
15	Suggestions for improvement (x_{15})	-0.039	0.058	-0.076	-0.671
16	Communication through mobile (x_{16})	0.083	0.050	0.181	1.654
17	No. of fragments (x_{17})	0.002	0.089	0.003	0.022
18	No. of stakeholder in different practices (x_{18})	-0.312	0.148	-0.497	-2.112
19	Total Production (x_{19})	-0.079	0.080	-0.113	-0.990
20	Family expenditure (x_{20})	-0.015	0.080	-0.020	-0.184
21	Energy consumption (x_{21})	-0.064	0.061	-0.121	-1.049
22	Nutrition and Health Status (x_{22})	-0.066	0.086	-0.140	-0.767

R square: 27.10%

The standard error of the estimate: 0.9080850

Table:7 presents the multiple regression analysis between Family Farm Income and 22 causal variables

It has been found that the 22 causal variables combined together have explained 57.03% of variance in the consequent variable Family Farm Income.

Subsequently, through stepwise regression analysis (Table no- 8) has been found that two variables, cropping intensity extent and constraint perception (x14) have been retained at the last step. Both cropping intensity and constraint perception have recorded significant contribution in improving Family Farm Income.

The higher is the cropping intensity, the better has been the Family Farm Income but it is also discernable that if crop combinations are not compatible, it may not improved Family Farm Income. On the contrary, if perception on constraints valid, the income has improved faster for the operating farms.

Table:8: Stepwise Regression Analysis: (y_2) Vs. 22 Causal Variables

Sl. No.	Variables	Reg. coeff. B	S.E. B	Beta	t value
1	Holding size (x6)	49963.384	10217.042	0.436	4.890

R square: 19.00%

The standard error of the estimate: 0.0974703

Table:8 presentsthe stepwise regressionanalysis where holding size (x_6) has been retained at the last step

When holding size goes higher, the productivity will turn increased and diverse. Hence, income and wage will go up.

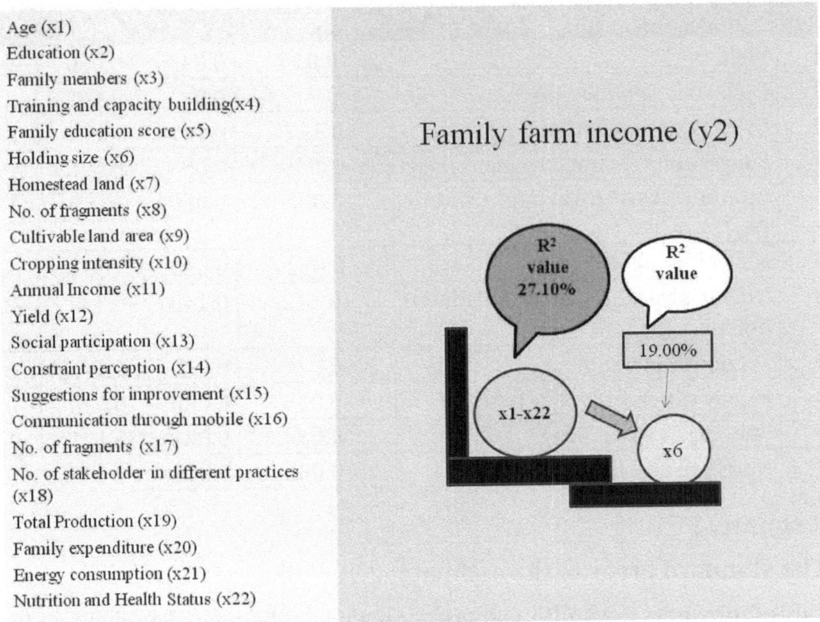

Model 4

Table 9: Path Analysis: Decomposition of Total Effect into Direct, Indirect and Residual Effect: Y_2

Sl. No.	*Variables*	*Total Effect*	*Direct Effect*	*Indirect Effect*	*Highest Indirect Effect*
1	Age (x_1)	0.149	0.088	0.061	0.181 (x_6)
2	Education (x_2)	0.152	0.103	0.049	0.116 (x_6)
3	Family members (x_3)	0.125	-0.495	0.620	0.308 (x_6)
4	Training and capacity building (x_4)	-0.029	0.012	-0.041	0.07 (x_3)
5	Family education score (x_5)	0.115	0.002	0.113	0.055 (x_{19})
6	Holding size (x_6)	0.438	0.570	-0.132	-0.268 (x_3)
7	Homestead land (x_7)	0.256	-0.120	0.376	0.263 (x_6)
8	No. of fragments (x_8)	0.134	-0.097	0.231	0.23 (x_6)
9	Cultivable land area (x_9)	0.205	0.050	0.155	0.323 (x_6)
10	Cropping intensity (x_{10})	-0.002	0.088	-0.090	-0.092 (x_6)
11	Annual Income (x_{11})	0.268	0.201	0.067	0.267 (x_6)
12	Yield (x_{12})	0.016	0.013	0.003	-0.044 (x_{21})
13	Social participation (x_{13})	0.055	0.002	0.053	0.042 (x_3)

Sl. No.	Variables	Total Effect	Direct Effect	Indirect Effect	Highest Indirect Effect
14	Constraint perception (x_{14})	0.015	-0.056	0.071	0.042 (x_{22})
15	Suggestions for improvement (x_{15})	0.153	0.177	-0.024	-0.073 (x_3)
16	Communication through mobile (x_{16})	-0.208	-0.092	-0.116	-0.181 (x_6)
17	No. of fragments (x_{17})	0.250	-0.139	0.389	0.336 (x_6)
18	No. of stakeholder in different practices (x_{18})	0.272	-0.137	0.409	0.316 (x_6)
19	Total Production (x_{19})	0.128	0.227	-0.099	-0.097 (x_3)
20	Family expenditure (x_{20})	-0.130	-0.168	0.038	0.051 (x_6)
21	Energy consumption (x_{21})	-0.271	-0.227	-0.044	0.160 (x_6)
22	Nutrition and Health Status (x_{22})	0.180	0.276	-0.096	-0.395 (x_3)

Residual effect:0.62707

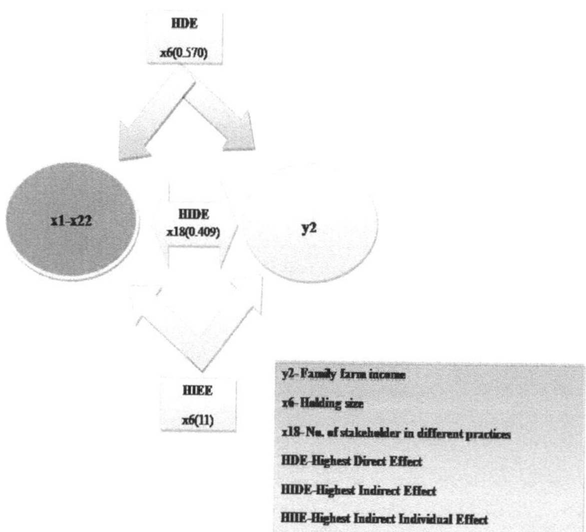

Model

Table:9 presents the Path analysis by decomposing the total effect into direct, indirect and residual effect.

Holding size has been recorded the highest substantive effect on farm income. So, it is clear that the holding size as resource endowment owned by the farmer has become the strong determinant to improve the farm income. It is further to infer that even today farm income is very much scale sensitive and goes better for the bigger size. The indirect effect, however, has been executed by the no. of stakeholders. The connectivity and interaction by a farmer with multistakeholder entities have contributed positively to make farmer successful enterprise.

The variable Holding size (x_6) has routed the highest indirect effect of as many as 13 exogenous variable to ultimately characterized the behaviour and performance the consequent variable i.e. Family Farm Income.

Table:10: Coefficient of Correlation (r): (y_3)vs. 22Independent Variables

Sl. No.	Independent Variables	'r' Value	Remarks
1	Age (x_1)	-0.075	
2	Education (x_2)	-0.029	
3	Family members (x_3)	-0.155	
4	Training and capacity building (x_4)	.316**	**
5	Family education score (x_5)	0.110	
6	Holding size (x_6)	-0.162	
7	Homestead land (x_7)	0.113	
8	No. of fragments (x_8)	-0.031	
9	Cultivable land area (x_9)	0.059	
10	Cropping intensity (x_{10})	-0.021	
11	Annual Income (x_{11})	-0.165	
12	Yield (x_{12})	.637**	**
13	Social participation (x_{13})	0.042	
14	Constraint perception (x_{14})	-0.094	
15	Suggestions for improvement (x_{15})	-0.129	
16	Communication through mobile (x_{16})	0.155	
17	No. of fragments (x_{17})	-0.105	
18	No. of stakeholder in different practices (x_{18})	-.259**	**
19	Total Production (x_{19})	-0.072	
20	Family expenditure (x_{20})	-0.146	
21	Energy consumption (x_{21})	0.147	
22	Nutrition and Health Status (x_{22})	-0.012	

**Correlation is significant at the 0.01 level

*Correlation is significant at the 0.05 level

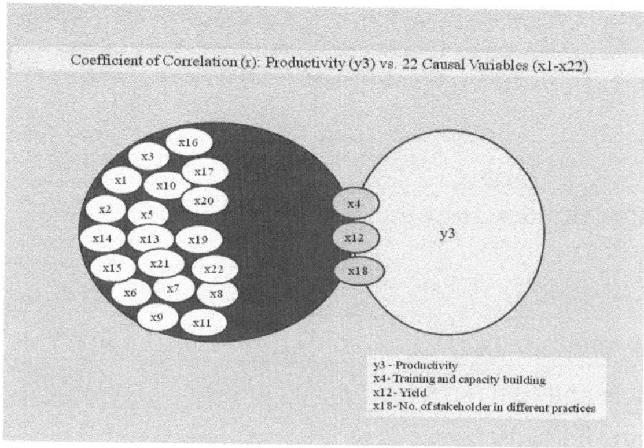

Model 5

Table:10 Presents the coefficient of correlation between productivity and 22 independent variables. Training and Capacity Building (x_4) has recorded positive and significant correlation. This indicates that intensive training has been instrumental in improving the productivity of farm.

Yield of crops (x_{12}) has collectively contributed to the overall productivity of farm and the relation is obvious.

No. of stakeholders (x_{18}) evinced that for leisure no. of stakeholder, the productivity has gone up.

Table:11.Multiple Regression Analysis: (y_3) Vs. 22 Causal Variables

Sl. No.	Variables	Reg. Coeff. B	S.E. B	Beta	t Value
1	Age (x_1)	0.076	0.092	0.087	0.820
2	Education (x_2)	0.071	0.080	0.102	0.895
3	Family members (x_3)	-0.364	0.132	-0.494	-2.762
4	Training and capacity building (x_4)	0.012	0.095	0.013	0.126
5	Family education score (x_5)	0.003	0.113	0.003	0.026
6	Holding size (x_6	0.568	0.184	0.571	3.094
7	Homestead land (x_7)	-0.145	0.147	-0.121	-0.986
8	No. of fragments (x_8)	-0.084	0.088	-0.098	-0.955
9	Cultivable land area (x_9)	0.062	0.187	0.048	0.332
10	Cropping intensity (x_{10})	0.053	0.066	0.087	0.808
11	Annual Income (x_{11})	0.217	0.222	0.203	0.979
12	Yield (x_{12})	0.008	0.065	0.012	0.119
13	Social participation (x_{13})	0.002	0.116	0.002	0.021
14	Constraint perception (x_{14})	-0.041	0.075	-0.056	-0.545
15	Suggestions for improvement (x_{15})	0.157	0.093	0.177	1.686

Sl. No.	Variables	Reg. Coeff. B	S.E. B	Beta	t Value
16	Communication through mobile (x_{16})	-0.073	0.081	-0.093	-0.910
17	No. of fragments (x_{17})	-0.153	0.142	-0.138	-1.071
18	No. of stakeholder in different practices (x_{18})	-0.150	0.237	-0.138	-0.634
19	Total Production (x_{19})	0.272	0.127	0.227	2.135
20	Family expenditure (x_{20})	-0.215	0.129	-0.168	-1.670
21	Energy consumption (x_{21})	-0.208	0.097	-0.228	-2.136
22	Nutrition and Health Status (x_{22})	0.224	0.137	0.276	1.632

R square: 37.30%

The standard error of the estimate: .1570197

Table-11presents the multiple regression analysis between productivity (y3) and 22 causal variables. All these 22 variables got together has contributed 37.30 percent of variants in the consequent variable-productivity.

Table:12.Stepwise Regression Analysis: (y_3) Vs. 22 Causal Variables

Sl. No.	Variables	Reg. Coeff. B	S.E. B	Beta	t value
1	Yield (x12)	0.856	0.101	0.626	8.505
2	Training and capacity building (x4)	899.126	257.177	0.262	3.496
3	Cropping intensity (x10)	-38.929	15.808	-0.185	-2.463

R square: 28.10%

The standard error of the estimate: 0.1590416

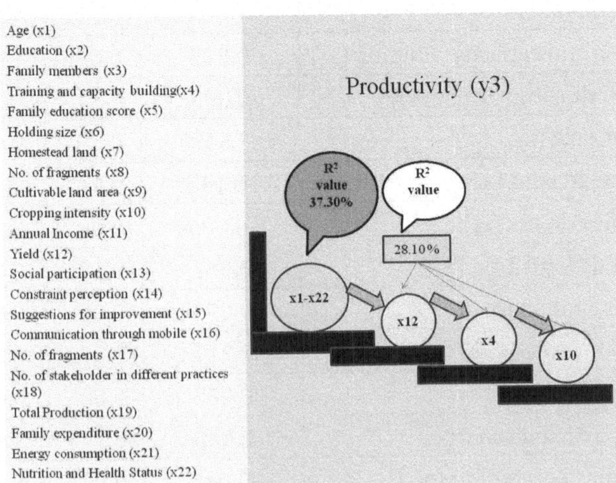

Model 6

Table:12 presents the regression analysis wherein the following variables, yield, training and capacity building and cropping intensity have been retained at the last step to substantially impact on Productivity.

Yield of a single crop is the strongest determinant of productivity. Training enhances capacity building in an individual and helps improve productive performance. Similarly, cropping intensity directly impacts on productivity by making every factor production perfectly orchestrating in any farming system.

Table:13.Path Analysis: Decomposition of Total Effect into Direct, Indirect and Residual Effect: Y3

Sl. No.	*Variables*	*Total Effect*	*Direct Effect*	*Indirect Effect*	*Highest Indirect Effect*
1	Age (x_1)	-0.075	-0.011	-0.064	-0.081 (x_6)
2	Education (x_2)	-0.029	-0.070	0.041	0.051 (x_6)
3	Family members (x_3)	-0.155	-0.097	-0.058	-0.138 (x_6)
4	Training and capacity building (x_4)	0.316	0.217	0.099	0.095 (x^{12})
5	Family education score (x_5)	0.110	0.015	0.095	0.039 (x_7)
6	Holding size (x_6)	-0.162	-0.254	0.092	0.089 (x_7)
7	Homestead land (x_7)	0.113	0.191	-0.078	-0.118 (x_6)
8	No. of fragments (x_8)	-0.031	0.052	-0.083	-0.103 (x_6)
9	Cultivable land area (x_9)	0.059	-0.012	0.071	-0.144 (x_6)
10	Cropping intensity (x_{10})	-0.021	-0.184	0.163	0.091 (x_{12})
11	Annual Income (x_{11})	-0.165	0.121	-0.286	-0.119 (x_6)
12	Yield (x_{12})	0.637	0.582	0.055	0.036 (x_4)
13	Social participation (x_{13})	0.042	-0.017	0.059	0.098 (x_{12})
14	Constraint perception (x_{14})	-0.094	-0.109	0.015	0.028 (x_{12})
15	Suggestions for improvement (x_{15})	-0.129	-0.140	0.011	0.049 (x_7)
16	Communication through mobile (x_{16})	0.155	-0.021	0.176	0.088 (x_{12})
17	No. of fragments (x_{17})	-0.105	0.130	-0.235	-0.15 (x_6)
18	No. of stakeholder in different practices (x_{18})	-0.259	-0.093	-0.166	-0.141 (x_6)
19	Total Production (x_{19})	-0.072	-0.132	0.060	0.048 (x_7)

Sl. No.	Variables	Total Effect	Direct Effect	Indirect Effect	Highest Indirect Effect
20	Family expenditure (x_{20})	-0.146	-0.037	-0.109	-0.118 (x_{12})
21	Energy consumption (x_{21})	0.147	0.017	0.130	0.112 (x_{12})
22	Nutrition and Health Status (x_{22})	-0.012	0.042	-0.054	-0.121 (x_6)

Residual effect: 0.44477

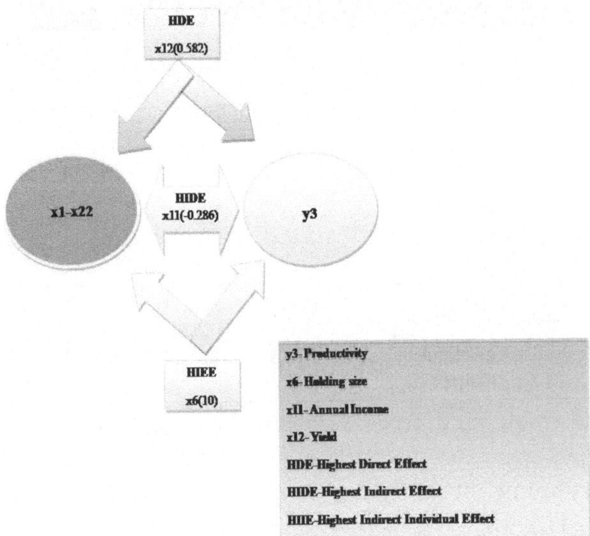

Model

Table-13 presents the path analysis by decomposing the total effect into direct, indirect and residual effect.

The variable yield of crops (x_{12}) has contributed the highest direct effect on the productivity, and this is for an obvious reason that when yield of individual crops increases, the productivity, of the farm as a whole improves.

The annual income as exerted the highest indirect effect on productivity. This may derive an element of doubt that whether the income from farm has proportionately been recycled into the overall process of farm entrepreneurship.

The variable-Holding size (x_6) has routed the highest indirect effect of as many as 10 causal variables to ultimately characterize the performance of the consequent variable, productivity. So in most cases, whatever may be the contributory factors, it needs to accompany the associate property of holding size.

Table:14.Coefficient of Correlation (r): Income per bigha (y_4) vs. 22Independent Variables

Sl. No.	Independent Variables	'r' Value	Remarks
1	Age (x_1)	.298**	**
2	Education (x_2)	-0.032	
3	Family members (x_3)	.339**	**
4	Training and capacity building (x_4)	-0.025	
5	Family education score (x_5)	-0.018	
6	Holding size (x_6)	.340**	**
7	Homestead land (x_7)	.273**	**
8	No. of fragments (x_8)	.211*	*
9	Cultivable land area (x_9)	-0.040	
10	Cropping intensity (x_{10})	0.153	
11	Annual Income (x_{11})	.766**	**
12	Yield (x_{12})	0.190	
13	Social participation (x_{13})	0.115	
14	Constraint perception (x_{14})	0.124	
15	Suggestions for improvement (x_{15})	0.113	
16	Communication through mobile (x_{16})	-0.100	
17	No. of fragments (x_{17})	0.073	
18	No. of stakeholder in different practices (x_{18})	.644**	**
19	Total Production (x_{19})	0.190	
20	Family expenditure (x_{20})	0.055	
21	Energy consumption (x_{21})	-0.006	
22	Nutrition and Health Status (x_{22})	.440**	**

**Correlation is significant at the 0.01 level

*Correlation is significant at the 0.05 level

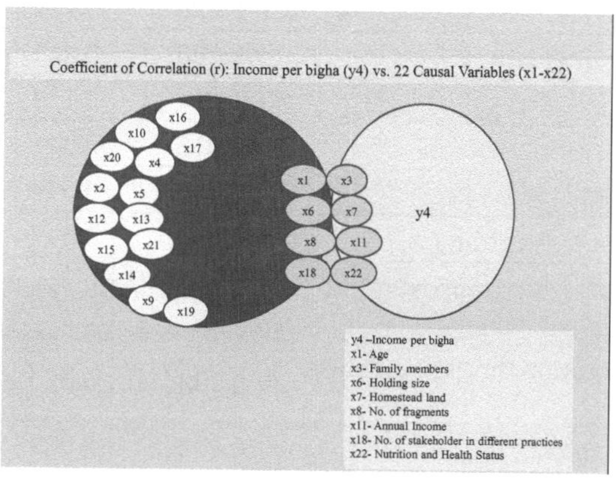

Model 7

Table-14 presents thecoefficient of correlationwithannual income per bigha and 22 exogenous variables.

The younger generation in rural people are withdrawing themselves from agriculture.

Family with higher number of members are engaged in different types of agricultural operations.

It is also found that holding size, homestead land, no. of fragments have generated more annual income.

It is interesting to note that the higher the family income, the more is the annual income.

The farms generating higher annual income are also engaging more number of stakeholders having better nutrition and health status. So, the farm production goes better.

Table:15. Multiple Regression Analysis: Income per bigha(y4) Vs. 22 Causal Variables

Sl. No.	*Variables*	*Reg. Coeff. B*	*S.E. B*	*Beta*	*t Value*
1	Age (x_1)	0.061	0.058	0.121	1.055
2	Education (x_2)	0.021	0.050	0.052	0.426
3	Family members (x_3)	0.037	0.082	0.087	0.450
4	Training and capacity building (x_4)	-0.026	0.060	-0.049	-0.443
5	Family education score (x_5)	0.018	0.071	0.032	0.253
6	Holding size (x_6)	0.248	0.115	0.430	2.160
7	Homestead land (x_7)	0.001	0.092	0.002	0.012
8	No. of fragments (x_8)	-0.024	0.055	-0.049	-0.441
9	Cultivable land area (x_9)	-0.228	0.117	-0.305	-1.951
10	Cropping intensity (x_{10})	-0.108	0.041	-0.306	-2.624
11	Annual Income (x_{11})	0.200	0.139	0.322	1.441
12	Yield (x_{12})	-0.018	0.041	-0.050	-0.451
13	Social participation (x_{13})	-0.112	0.073	-0.165	-1.548
14	Constraint perception (x_{14})	0.117	0.047	0.277	2.496
15	Suggestions for improvement (x_{15})	-0.039	0.058	-0.076	-0.671
16	Communication through mobile (x_{16})	0.083	0.050	0.181	1.654
17	No. of fragments (x_{17})	0.002	0.089	0.003	0.022
18	No. of stakeholder in different practices (x_{18})	-0.312	0.148	-0.497	-2.112

Sl. No.	*Variables*	*Reg. Coeff. B*	*S.E. B*	*Beta*	*t Value*
19	Total Production (x_{19})	-0.079	0.080	-0.113	-0.990
20	Family expenditure (x_{20})	-0.015	0.080	-0.020	-0.184
21	Energy consumption (x_{21})	-0.064	0.061	-0.121	-1.049
22	Nutrition and Health Status (x_{22})	-0.066	0.086	-0.140	-0.767

R square:27.10%

The standard error of the estimate: 0.980850

Table-15 presents the multiple regression analysis between per bigha (y_4) income and 22 independent variables

It has been found that age (x_1) has contributed positively to improve in farm income.So,in this case,experienced farmers have been successful to increase farm income.So,here in this case,experienced farmers have been successful to increase farm income than the younger farmers.

Family members (x_3) has recorded a positive correlation with farm income which implies that higher size of family has contributed positively towards farm improving farm income.

Homestead land (x_7) also has contributed to farm income in a positive ways. So, higher is the homestead land, better has been the farm income.

No. of fragments (x_8) has contributed positively towards farm income. This implies that more number of fragments has gone positively to escalate farm income by indulging higher crop diversity.

Family income has rightly contributed to the escalation of farm income. The farmers generating higher farm income have been found to engage higher no. of stakeholders involved in different practices (x_{18}).

Families with higher health and nutrition status have also been found to generate higher farm income.

Table:16. Stepwise Regression Analysis:Income per bigha(y_4) Vs. 22 Causal Variables

Sl.No.	*Variables*	*Reg. coef.B*	*S.E. B*	*Beta*	*t value*
1	Annual Income (x_{11})	1.002	0.006	0.987	164.066
2	Total Production (x_{19})	0.154	0.020	0.042	7.597
3	Homestead land (x_7)	3434.633	1062.786	0.019	3.232
4	Family members (x_3)	-4384.334	2170.734	-0.012	-2.020

R square: 15.50%

The standard error of the estimate: 0.974703

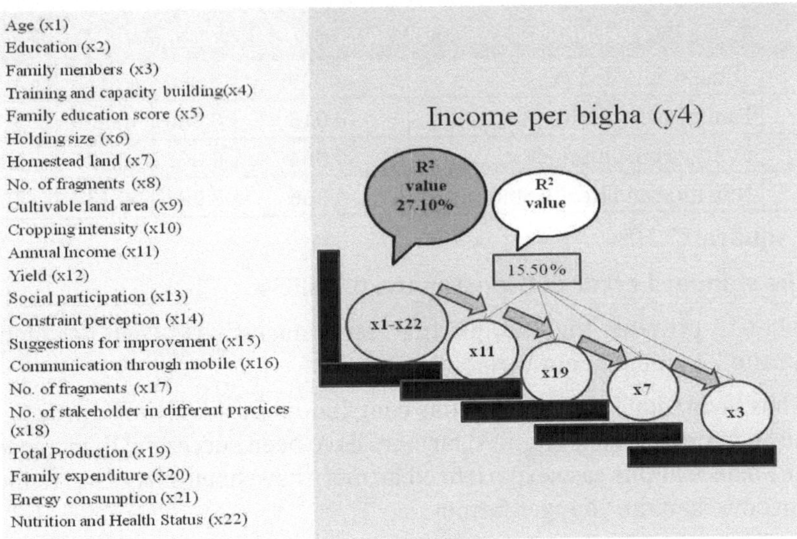

Model 8

Table-16 presents the stepwise regression analysis whereinthe following variables; annual income, total production, and homestead land, family membershave been retained at the last step.

Annual income is the direct estimator of farm income, that's why it has got a discernible impact on income per bigha. The farm income has strongly been predicted by total production and homestead land and the causal relation is obvious. Family members have helped here to generate collective and productive decision. Additionally, big size family means availability of family skills and labour by default.

Table: 17. Path Analysis: Decomposition of Total Effect into Direct, Indirect and Residual Effect: Income per bigha (y_4)

Sl. No.	Variables	Total Effect	Direct Effect	Indirect Effect	Highest Indirect Effect
1	Age (x_1)	0.298	0.081	0.217	0.218 (x_{11})
2	Education (x_2)	-0.032	0.105	-0.137	-0.051 (x_{11})
3	Family members (x_3)	0.339	-0.311	0.650	0.331 (x_{11})
4	Training and capacity building (x_4)	-0.025	-0.006	-0.019	-0.124 (x_{11})
5	Family education score (x_5)	-0.018	-0.109	0.091	0.039 (x_2)
6	Holding size (x_6)	0.340	-0.074	0.414	0.366 (x_{11})
7	Homestead land (x_7)	0.273	-0.031	0.304	0.265 (x_{11})

Sl. No.	Variables	Total Effect	Direct Effect	Indirect Effect	Highest Indirect Effect
8	No. of fragments (x_8)	0.211	0.051	0.160	0.177 (x_{11})
9	Cultivable land area (x_9)	-0.040	0.110	-0.150	-0.068 (x_3)
10	Cropping intensity (x_{10})	0.153	0.111	0.042	-0.036 (x_9)
11	Annual Income (x_{11})	0.766	0.782	-0.016	-0.132 (x_3)
12	Yield (x_{12})	0.190	0.185	0.005	-0.078 (x_{11})
13	Social participation (x_{13})	0.115	-0.003	0.118	0.031 (x_{12})
14	Constraint perception (x_{14})	0.124	-0.074	0.198	0.104 (x_{11})
15	Suggestions for improvement (x_{15})	0.113	0.040	0.073	0.089 (x_{11})
16	Communication through mobile (x_{16})	-0.100	0.024	-0.124	-0.191 (x_{11})
17	No. of fragments (x_{17})	0.073	-0.065	0.138	0.122 (x_{11})
18	No. of stakeholder in different practices (x_{18})	0.644	0.043	0.601	0.688 (x_{11})
19	Total Production (x_{19})	0.190	0.142	0.048	0.165 (x_{11})
20	Family expenditure (x_{20})	0.055	-0.035	0.090	0.103 (x_{11})
21	Energy consumption (x_{21})	-0.006	0.052	-0.058	-0.134 (x_{11})
22	Nutrition and Health Status (x_{22})	0.440	0.346	0.094	0.284 (x_{11})

Residual effect: 0.26599

Table:17 presents the path analysis through the decomposition of total effect into direct, indirect and residual effect

The variable, family members, (x3) have exerted the highest indirect effect.It means that the size of the family is an important determinant exerting intensive associating property to settle the family income.

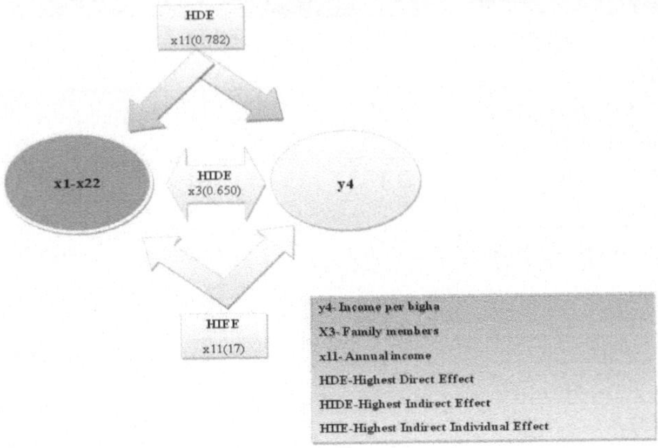

Model

On a simpler note,it is to infer that the farm income has been impacted by both becoming potential consumer and potential contributor in the form of family labour.

The variable annual income (x_{11}) has got a direct and substantive effect impacts on farm income.The higher is the family income, the better and stronger are the investment ability which results into higher farm income.

An assuredand higher family income ensure adequate and proper time of application to exert the proficiency of input management which reflects in the farm of higher yield and better income.

It implies that combination of 22 exogenous variable could not explain 26.59 per cent of variance in the consequent variable,Income per bigha(y_4).

Table:18.Coefficient of Correlation (r): (y_5) vs. 22 Independent Variables

Sl. No.	*Independent Variables*	*'r' Value*	*Remarks*
1	Age (x_1)	.259**	**
2	Education (x_2)	-0.020	
3	Family members (x_3)	.414**	**
4	Training and capacity building (x_4)	-0.172	
5	Family education score (x_5)	-0.001	
6	Holding size (x_6)	.521**	**
7	Homestead land (x_7)	.333**	**
8	No. of fragments (x_8)	.295**	**
9	Cultivable land area (x_9)	-0.046	
10	Cropping intensity (x_{10})	-0.015	
11	Annual Income (x_{11})	.855**	**

Sl. No.	Independent Variables	'r' Value	Remarks
12	Yield (x_{12})	-0.153	
13	Social participation (x_{13})	0.009	
14	Constraint perception (x_{14})	0.115	
15	Suggestions for improvement (x_{15})	0.121	
16	Communication through mobile (x_{16})	-.242*	*
17	No. of fragments (x_{17})	0.181	
18	No. of stakeholder in different practices (x_{18})	.897**	**
19	Total Production (x_{19})	.204*	*
20	Family expenditure (x_{20})	0.138	
21	Energy consumption (x_{21})	-0.189	
22	Nutrition and Health Status (x_{22})	.311**	**

**Correlation is significant at the 0.01 level

*Correlation is significant at the 0.05 level

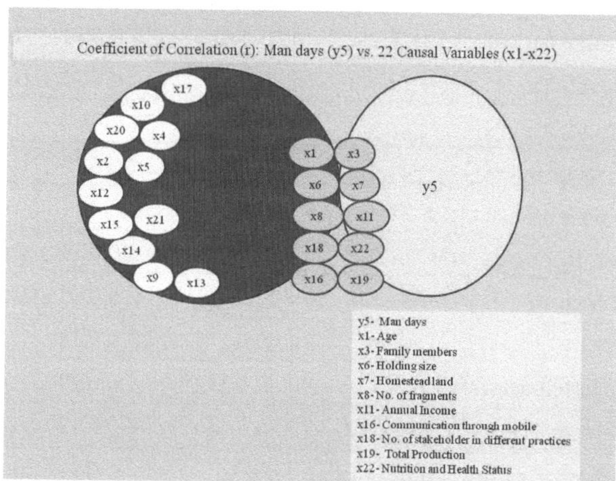

Model 9

Table–18 presents the coefficient of correlation between man days (y_5) and 22 independent variables.

It has been found that farmers with higher age echelon have generated more man days than the younger ones.The other side of the fact is that young generation rural people are increasingly withdrawing themselves from agriculture. This will throw a serious challenge to productivity issues in Indian agriculture.

Family with higher number of members are engaged more in agricultural operations.

It is also found that farmers with higher holding size, homestead land,higher number of fragmented land holdings have generated more man days.

It is interesting to note that higher fragmentation has invited generation of higher volume of man days. With move number of fragments, it will take more time and mandays to perform the same operation.

It is elicited that those who are generating more man days they are prone to interpersonal communication than to access mobile telephony.

The farms generating higher man days are also engaging more and diverse stakeholders having higher nutrition and better health status and in the end,their farm production goes better.

Table:19.Multiple Regression Analysis: (y_5) Vs. 22 Causal Variables

Sl. No.	Variables	Reg. Coef. B	S.E. B	Beta	t Value
1	Age (x_1)	-0.007	0.040	-0.010	-0.187
2	Education (x_2)	-0.029	0.035	-0.046	-0.826
3	Family members (x_3)	0.046	0.057	0.069	0.796
4	Training and capacity building (x_4)	-0.024	0.041	-0.029	-0.578
5	Family education score (x_5)	0.056	0.049	0.065	1.140
6	Holding size (x_6)	0.174	0.080	0.194	2.183
7	Homestead land (x_7)	-0.002	0.064	-0.002	-0.029
8	No. of fragments (x_8)	0.076	0.038	0.098	1.974
9	Cultivable land area (x_9)	-0.126	0.081	-0.109	-1.555
10	Cropping intensity (x_{10})	-9.039E-05	0.029	0.000	-0.003
11	Annual Income (x_{11})	0.180	0.096	0.186	1.863
12	Yield (x_{12})	0.012	0.028	0.022	0.437
13	Social participation (x_{13})	-0.007	0.050	-0.007	-0.146
14	Constraint perception (x_{14})	-0.013	0.033	-0.020	-0.396
15	Suggestions for improvement (x_{15})	0.036	0.041	0.045	0.891
16	Communication through mobile (x_{16})	0.053	0.035	0.074	1.502
17	No. of fragments (x_{17})	-0.073	0.062	-0.073	-1.184
18	No. of stakeholder in different practices (x_{18})	0.660	0.103	0.675	6.409
19	Total Production (x_{19})	0.014	0.055	0.013	0.246
20	Family expenditure (x_{20})	0.020	0.056	0.017	0.361
21	Energy consumption (x_{21})	0.050	0.042	0.061	1.185
22	Nutrition and Health Status (x_{22})	-0.092	0.060	-0.126	-1.549

R square: 85.40%

The standard error of the estimate: 0.682517

Table-19 presents the multiple regression analysis to depict causal correlation between man days (y_5) and 22 causal variables

These 22 causal variables have collectively contributed to explain 85.40% variance in the consequent variable, man days (y_5).

This implies that the combination of all these 22 causal variables has explained 85.40 per cent ion the variance in the consequent variable, man days (y_5).

Table 20. Stepwise Regression Analysis: (y_5) Vs. 22 Causal Variables

Sl.No.	*Variables*	*Reg. coef.B*	*S.E. B*	*Beta*	*t value*
1	No. of stakeholder in different practices (x_{18})	0.627	0.085	0.641	7.364
2	Annual Income (x_{11})	0.280	0.084	0.291	3.338

R square: 82.40%

The standard error of the estimate: 0.671896

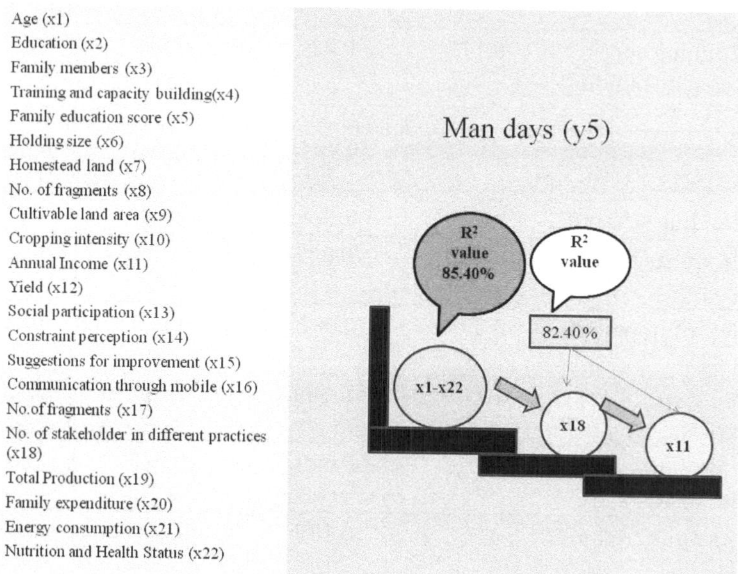

Model 10

Table–20 presents the stepwise regression analysis for the screening out of dominant causal variables.

It has been found that two variables:-no. of stakeholders involved in different practices (x_{18}) and annual income (x_{11}) have been retained at the last step in the process of backward elimination of stepwise regression analysis.

It indicates that man days have been generated the most wherein no. of stakeholders (x_{18}) is on a higher side.With the diversification of farming through inclusion of elite enterprises like fishery,poultry,horticulture alongside the main field crop enterprise, the no. of man days has been increased.

Farmers with higher annual income have been found to generate higher intensity of man days.

Involvement of diverse stakeholders, crop diversification and higher size of holding has been found to generate a combined and positive effect on generation of both high intensity of man days and productivity as well.

Table: 21.Path Analysis: Decomposition of Total Effect into Direct, Indirect and Residual Effect: Y5

Sl. No.	Variables	Total Effect	Direct Effect	Indirect Effect	Highest Indirect Effect
1	Age (x_1)	0.259	-0.010	0.269	0.193 (x_{18})
2	Education (x_2)	-0.020	-0.047	0.027	0.039 (x_6)
3	Family members (x_3)	0.414	0.070	0.344	0.3 (x_{18})
4	Training and capacity building (x_4)	-0.172	-0.027	-0.145	-0.123 (x_{18})
5	Family education score (x_5)	-0.001	0.064	-0.065	-0.03 (x_9)
6	Holding size (x_6)	0.521	0.191	0.330	0.374 (x_{18})
7	Homestead land (x_7)	0.333	-0.002	0.335	0.236 (x_{18})
8	No. of fragments (x_8)	0.295	0.097	0.198	0.165 (x_{18})
9	Cultivable land area (x_9)	-0.046	-0.105	0.059	0.108 (x_6)
10	Cropping intensity (x_{10})	-0.015	0.000	-0.015	0.034 (x_9)
11	Annual Income (x_{11})	0.855	0.189	0.666	0.593 (x_{18})
12	Yield (x_{12})	-0.153	0.021	-0.174	-0.139 (x_{18})
13	Social participation (x_{13})	0.009	-0.006	0.015	-0.017 (x_{18})
14	Constraint perception (x_{14})	0.115	-0.019	0.134	0.108 (x_{18})
15	Suggestions for improvement (x_{15})	0.121	0.046	0.075	0.076 (x_{18})

Sl. No.	*Variables*	*Total Effect*	*Direct Effect*	*Indirect Effect*	*Highest Indirect Effect*
16	Communication through mobile (x_{16})	-0.242	0.072	-0.314	-0.215 (x_{18})
17	No. of fragments (x_{17})	0.181	-0.073	0.254	0.171 (x_{18})
18	No. of stakeholder in different practices (x_{18})	0.897	0.674	0.223	0.166 (x_{11})
19	Total Production (x_{19})	0.204	0.013	0.191	0.114 (x_{18})
20	Family expenditure (x_{20})	0.138	0.017	0.121	0.097 (x_{18})
21	Energy consumption (x_{21})	-0.189	0.060	-0.249	-0.174 (x_{18})
22	Nutrition and Health Status (x_{22})	0.311	-0.127	0.438	0.25 (x_{18})

Residual effect: 0.14643

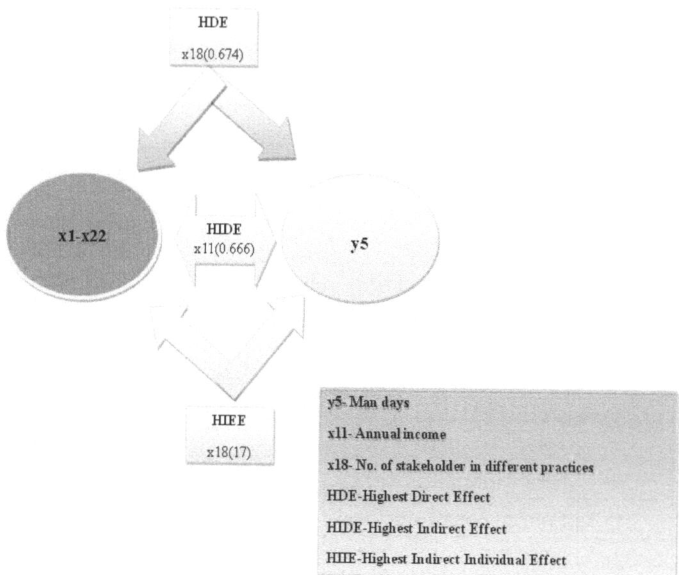

Model

Table-21 presents the path analysis through the decomposition of total effect into direct, indirect and residual effect

It has been found that no. of stakeholders involved in different practices (x_{18}) has exerted the highest direct effect and annual income,on the other hand,has exerted the highest indirect effect. Again,the variable, no. of stakeholders has been involved in different practices has exerted the highest associative impact by recording the highest frequency of indirect effect.

It has been found that the variable, no. of stakeholders (x_{18}), has enrooted on man days (y5) the highest indirect effect on as many as 16 exogenous variables to ultimately impact on the consequent variable man days (y_5).

It implies that the combination of these 2 exogenous variables could not explain 14.64 percent of variance in the consequent variable,man days (y_5).

Table: 22.Coefficient of Correlation (r): (y_6) vs. 22Independent Variables

Sl. No.	*Independent Variables*	*'r' Value*	*Remarks*
1	Age (x_1)	-0.149	
2	Education (x_2)	-0.042	
3	Family members (x_3)	-0.115	
4	Training and capacity building (x_4)	-0.006	
5	Family education score (x_5)	-0.043	
6	Holding size (x_6)	-0.099	
7	Homestead land (x_7)	-0.013	
8	No. of fragments (x_8)	0.174	
9	Cultivable land area (x_9)	-0.178	
10	Cropping intensity (x_{10})	-0.068	
11	Annual Income (x_{11})	0.001	
12	Yield (x_{12})	0.182	
13	Social participation (x_{13})	-0.057	
14	Constraint perception (x_{14})	-0.011	
15	Suggestions for improvement (x_{15})	-0.062	
16	Communication through mobile (x_{16})	-.216*	*
17	No. of fragments (x_{17})	-0.133	
18	No. of stakeholder in different practices (x_{18})	-0.028	
19	Total Production (x_{19})	-0.135	
20	Family expenditure (x_{20})	-0.046	
21	Energy consumption (x_{21})	-0.073	
22	Nutrition and Health Status (x_{22})	-0.077	

**Correlation is significant at the 0.01 level

*Correlation is significant at the 0.05 level

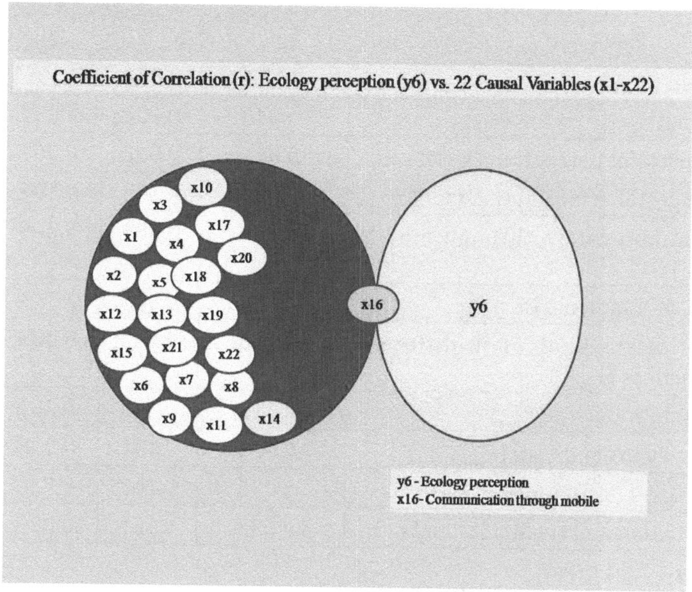

Model

Table 22presents the coefficient of correlation between ecology perception and 22 independent variables

It has been found that communication through mobile has recorded significant and negative correlation with ecology perception.

Mobile communication,by it's nature of impact, has recorded a negative correlation with ecology perception. Prolonged and unscientific use of mobile telephone has got an adverse effect on health and surrounding ecosystem as well.

Table: 23.Multiple Regression Analysis: (y_6) Vs. 22 Causal Variables

Sl. No.	Variables	Reg. Coef. B	S.E. B	Beta	t Value
1	Age (x_1)	-0.071	0.101	-0.079	-0.698
2	Education (x_2)	-0.025	0.088	-0.034	-0.282
3	Family members (x_3)	-0.055	0.145	-0.072	-0.380
4	Training and capacity building (x_4)	0.021	0.105	0.022	0.204
5	Family education score (x_5)	0.037	0.124	0.037	0.294
6	Holding size (x_6)	-0.076	0.202	-0.074	-0.378
7	Homestead land (x_7)	0.029	0.162	0.024	0.182
8	No. of fragments (x_8)	0.245	0.097	0.274	2.527
9	Cultivable land area (x_9)	-0.406	0.206	-0.304	-1.977
10	Cropping intensity (x_{10})	-0.116	0.072	-0.184	-1.608
11	Annual Income (x_{11})	0.136	0.244	0.123	0.559

Sl. No.	Variables	Reg. Coef. B	S.E. B	Beta	t Value
12	Yield (x_{12})	0.173	0.072	0.264	2.402
13	Social participation (x_{13})	-0.045	0.128	-0.037	-0.354
14	Constraint perception (x_{14})	-0.051	0.082	-0.067	-0.615
15	Suggestions for improvement (x_{15})	-0.099	0.103	-0.107	-0.965
16	Communication through mobile (x_{16})	-0.221	0.089	-0.268	-2.490
17	No. of fragments (x_{17})	0.044	0.156	0.038	0.278
18	No. of stakeholder in different practices (x_{18})	-0.100	0.260	-0.089	-0.385
19	Total Production (x_{19})	-0.313	0.140	-0.252	-2.239
20	Family expenditure (x_{20})	-0.035	0.141	-0.026	-0.249
21	Energy consumption (x_{21})	-0.036	0.107	-0.038	-0.337
22	Nutrition and Health Status (x_{22})	-0.033	0.151	-0.039	-0.216

R square: 29.50%

The standard error of the estimate: 0.1725642

Table- 23 presents the multiple regression analysis between ecology perception and 22 causal variables

These 22 causal variables have collectively contributed to explain 29.50 percent variance in the consequent variable, ecology perception (y_6).

Table:24.Stepwise Regression Analysis: (y_6) Vs. 22 Causal Variables

Sl.No.	Variables	Reg.coef.B	S.E. B	Beta	t value
1	Communication through mobile (x_{16})	-0.196	0.075	-0.239	-2.631
2	Yield (x_{12})	0.181	0.059	0.277	3.069
3	Cultivable land area (x_9)	-0.454	0.129	-0.339	-3.507
4	No. of fragments (x_8)	0.200	0.081	0.224	2.471
5	Total Production (x_{19})	-0.287	0.113	-0.230	-2.540

R square: 27.40%

The standard error of the estimate:0.1631220

Table- 24presents the stepwise regression analysis and it has been found that the following variables –communication through mobile, yield, cultivable land area, no. of fragments, total production have been retained at the last step.

Yield (x_{12}) is an ecological function so it has got a substantive contribution perception.

Cultivable land (x_9) has a natural resource is well integrated to ecology. The nature and approach to management of the land has got correlates to energy management, methane emission, and water economy and biodiversity conservation.

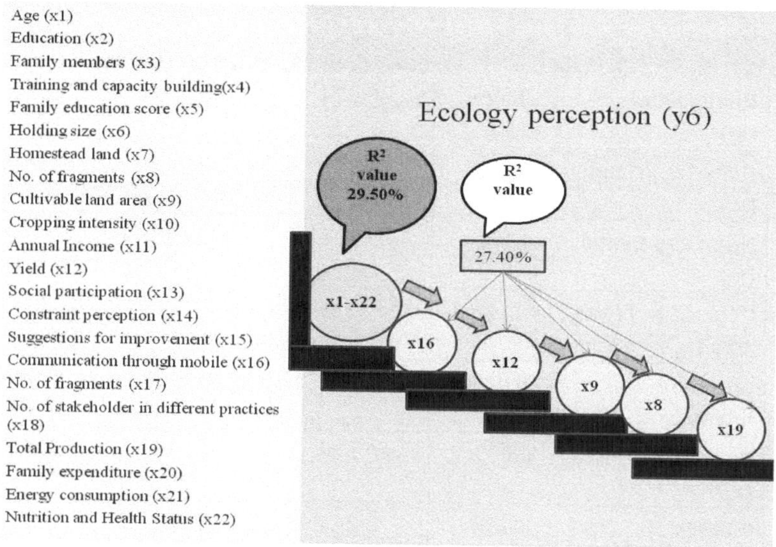

Model

The alarming feature of Indian operational holdings are it's marginal characters and keeps fragmenting in a response to disintegration of family. High rate of fragmentation makes operational holding both energy and resource prodigal.

Since yield is an ecological function,production by default is interacting with ecological function as well. If resilience of the agro-ecosystem is lost to boost up agricultural productivity,the resource base will turn fragile and the resilience will be lost. So,better ecology perception contributes to better land water and energy management and subsequently the restoration of resilience and productivity both.

Table:25.Path Analysis: Decomposition of Total Effect into Direct, Indirect and Residual Effect: Y6

Sl. No.	*Variables*	*Total Effect*	*Direct Effect*	*Indirect Effect*	*Highest Indirect Effect*
1	Age (x_1)	-0.149	-0.079	-0.070	0.034 (x_{11})
2	Education (x_2)	-0.042	-0.035	-0.007	-0.083 (x_9)
3	Family members (x_3)	-0.115	-0.073	-0.042	0.083 (x_8)
4	Training and capacity building (x_4)	-0.006	0.023	-0.029	-0.067 (x_{16})
5	Family education score (x_5)	-0.043	0.037	-0.080	-0.086 (x_9)

Sl. No.	Variables	Total Effect	Direct Effect	Indirect Effect	Highest Indirect Effect
6	Holding size (x_6)	-0.099	-0.074	-0.025	-0.173 (x_9)
7	Homestead land (x_7)	-0.013	0.024	-0.037	-0.063 (x_{19})
8	No. of fragments (x_8)	0.174	0.274	-0.100	-0.066 (x_9)
9	Cultivable land area (x_9)	-0.178	-0.305	0.127	0.06 (x_{10})
10	Cropping intensity (x_{10})	-0.068	-0.185	0.117	0.098 (x_9)
11	Annual Income (x_{11})	0.001	0.119	-0.118	-0.075 (x_{18})
12	Yield (x_{12})	0.182	0.264	-0.082	-0.041 (x_{16})
13	Social participation (x_{13})	-0.057	-0.037	-0.020	-0.054 (x_{16})
14	Constraint perception (x_{14})	-0.011	-0.068	0.057	0.03 (x_9)
15	Suggestions for improvement (x_{15})	-0.062	-0.108	0.046	0.04 (x_{19})
16	Communication through mobile (x_{16})	-0.216	-0.268	0.052	0.04 (x_{12})
17	No. of fragments (x_{17})	-0.133	0.038	-0.171	-0.143 (x_9)
18	No. of stakeholder in different practices (x_{18})	-0.028	-0.085	0.057	0.105 (x_{11})
19	Total Production (x_{19})	-0.135	-0.252	0.117	0.037 (x_{10})
20	Family expenditure (x_{20})	-0.046	-0.027	-0.019	-0.053 (x_{12})
21	Energy consumption (x_{21})	-0.073	-0.038	-0.035	-0.08 (x_8)
22	Nutrition and Health Status (x_{22})	-0.077	-0.037	-0.040	0.073 (x_8)

Residual effect:0.70482

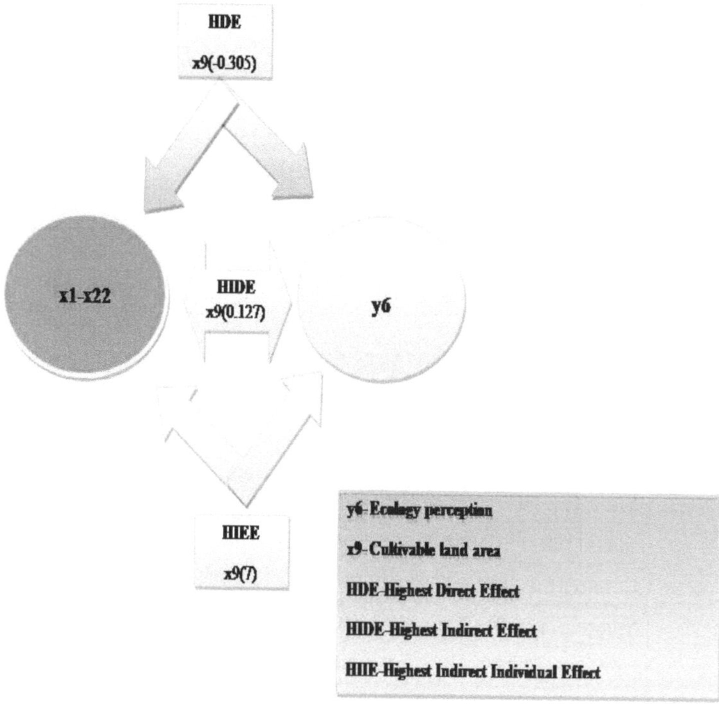

Model

Table- 25 presents the path analysis by decomposing the total effect into direct, indirect and residual effect.

No. of fragments (x_{17}) has contributed the substantive impact of ecological perception. Since,unabated fragmentation keeps on contributing to energy and resource prodigal nature of holdings, and, thereby, generating a deleterious impact on ecological resilience. This has generated a strong impact on ecology perception as well.

Table: 26.Canonical Covariate Analysis: Interaction between Left Side Dependent and Right Side Independent Variables

Dependent Variable	*Coefficients*
Average wage/day (y_1)	0.188
Family farm income (y_2)	0.222
Productivity (y_3)	0.085
Income per bigha(y_4)	-0.058
Man days (y_5)	-0.379
Ecology perception (y_6)	0.512

Independent Variable	Coefficients
Age (x_1)	0.303
Education (x_2)	0.094
Family members (x_3)	0.299
Training and capacity building (x_4)	-0.223
Family education score (x_5)	0.446
Holding size (x_6)	-0.209
Homestead land (x_7)	0.339
No. of fragments (x_8)	0.272
Cultivable land area (x_9)	0.135
Cropping intensity (x_{10})	0.421
Annual income (x_{11})	-0.539
Yield (x_{12})	-0.114
Social participation (x_{13})	0.050
Constraint perception (x_{14})	0.229
Suggestions for improvement (x_{15})	0.233
Communication through mobile (x_{16})	0.346
No. of fragments (x_{17})	0.135
No. of stakeholder in different practices (x_{18})	0.779
Total production (x_{19})	0.089
Family expenditure (x_{20})	0.313
Energy consumption (x_{21})	0.674
Nutrition and health status (x_{22})	-0.186

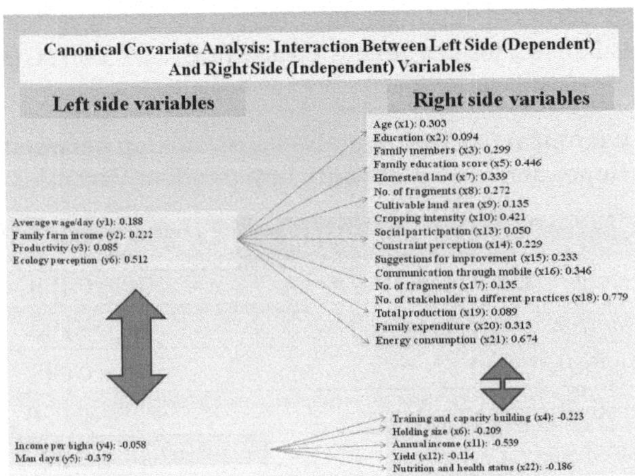

Model

Table: 26presents the Canonical Covariate Analysis wherein the whole set of variables has been divided into left side (y) and right side (x) variably.

The following conglomeration of y variables, $(y_1),(y_2),(y_3),(y_6)$,is moving together and these conglomeration of variables have selectively been interacting with the following sets of variables $(x_1,x_2,x_3,x_4,x_5,x_6,x_7,x_8,x_9,x_{10},x_{11},x_{12},x_{13},x_{14},x_{15},x_{16},x_{17},x_{18},x_{19},x_{20},x_{21},x_{22})$.So,an isochronous change in wage (y_1), income (y_2), productivity (y_3) and ecology perception (y_6) merits a concurrent contribution from the following exogenous variables.

Similarly,the other two dependent variables, farm income and man days, are moving together. The lion's share of cost in farming is incurred in the form of man days. So,income from farm and man days are moving together and this relation and interaction, a concurrent contribution, is coming from the following x variables.

The intra mode of interaction is based on cross loading. Here, in this Canonical Covariate Analysis, the y-set of variables has been regressed in terms of x-set of variables and on the contrary,the x-set of variables has been regressed in terms of variables.

Table: 27.Factor Analysis: Strategic Conglomeration of Variables into Factor

SL. No.	Variables	Factor loading	% of variance	Cumulative %	Factor renamed
1	Annual income (x_{11}) No. of stakeholder in different practices (x_{18}) Income per bigha (y_4) Man days (y_5)	0.934 0.885 0.811 0.895	20.047	20.047	Farm economy
2	Holding size (x_6) Cultivable land area (x_9) No. of fragments (x_{17})	0.710 0.777 0.785	9.081	29.129	Resource character
3	Yield (x_{12}) Productivity (y_3)	0.827 0.876	8.111	37.239	Productive function
4	Family members (x_3) Nutrition and health status (x_{22})	0.841 0.856	6.391	43.631	Family Health
5	Age (x_1) Education (x_2) Family education score (x_5)	-0.526 0.790 0.636	5.632	49.263	Educational capability
6	Training and capacity building (x_4) Cropping intensity (x_{10}) Suggestions for improvement (x_{15}) Total production (x_{19})	0.451 0.531 0.429 -0.748	5.386	54.649	Farm capacity

SL. No.	Variables	Factor loading	% of variance	Cumulative %	Factor renamed
7	No. of fragments (x_8) Communication through mobile (x_{16}) Ecology perception (y_6)	-0.466 0.553 -0.762	4.804	59.452	Ecological proficiency
8	Homestead land (x_7) Family expenditure (x_{20}) Family farm income (y_2) Energy consumption (x_{21})	0.412 -0.705 0.486 -0.554	4.509	63.961	Homestead dynamics
9	Average wage/day (y_1)	0.812	4.272	68.233	
10	Social participation (x_{13}) Constraint perception (x_{14})	0.788 0.572	3.704	71.937	Social ecology

Table- 27 presents the factor analysis wherein 28 variables have been conglomerated into 10 discernible factors

Factor Analysis: Conglomeration of 28 Explanatory Variables Into Sub Factors

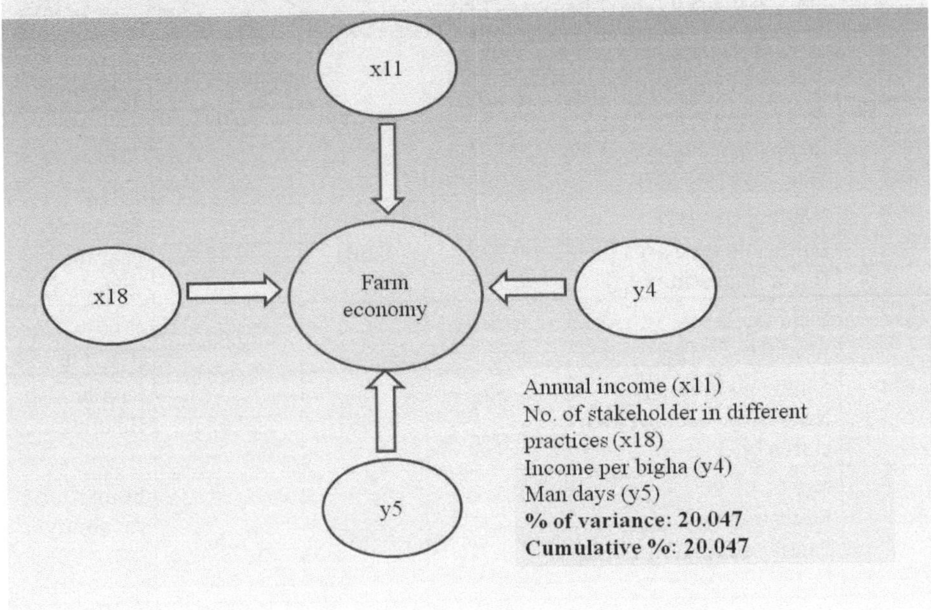

Model

Factor 1 has accommodated four homogenous variables (estimated through factor loading) to form factor1.This factor has contributed 20.05 per cent of variance and has been renamed as farm economy.

Factor Analysis: Conglomeration of28 Explanatory Variables Into Sub Factors

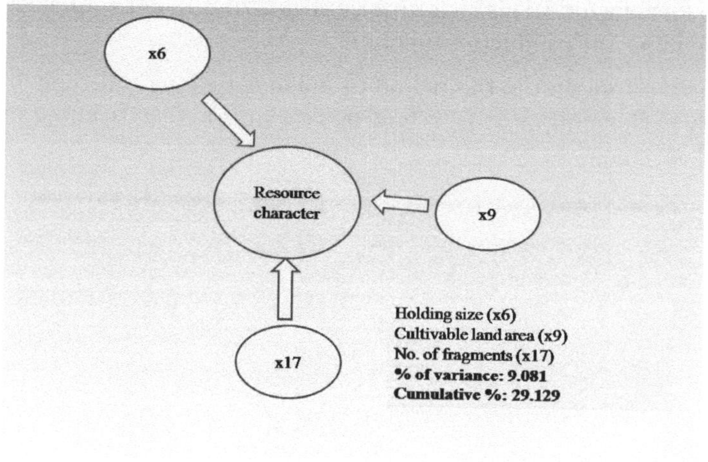

Model

Farm economy as a factor lays the foundation for the entire dynamics of interaction comprising of income, productivity, livelihood and management at large.

Factor 2 has accommodated three homogenous variables (estimated through factor loading) to form factor 2.This factor has contributed 9.09 per cent of variance and has been renamed as resource character. The complex and non-linear interactions amounts in between resource variables under the factors have contributed to the income level, income, productivity and management to the entire production regime and it's response to market.

Factor Analysis: Conglomeration of28 Explanatory Variables Into Sub Factors

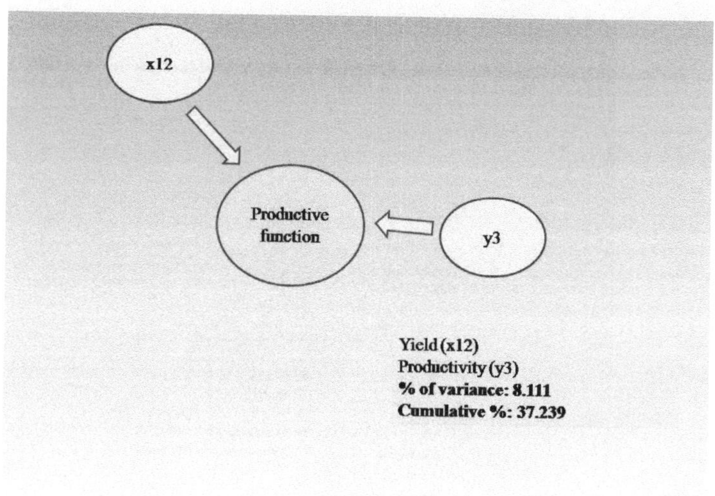

Factor

Factor 3 has accommodated two operating variables (estimated through factor loading) to form factor 3.This factor has contributed 8.12 per cent of variance and has been renamed as productive function.

Productive function is the prime to make a system and their interaction, successfully performing and that is how this variable has contributed significantly to the total variance.

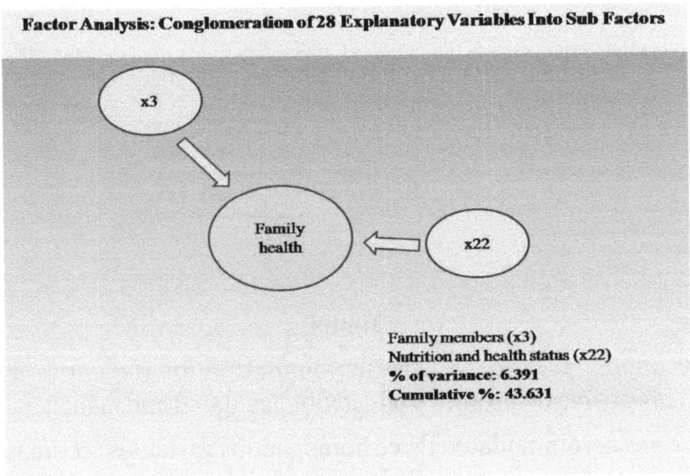

Model

Factor 4 has accommodated two functional variables (estimated through factor loading) to form factor 4.This factor has contributed 6.40 per cent of variance and has been renamed as family health.

Health and nutrition of farmers, engaged in innumerable farm operations, supports the productive performance of a farm, small, big or medium.

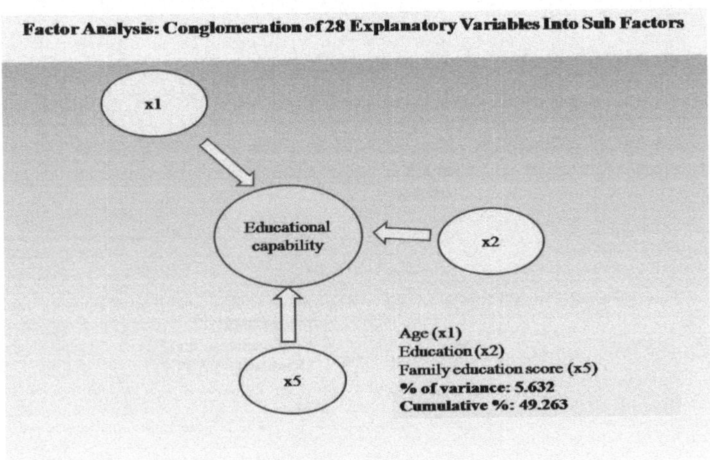

Model

Factor 5 has accommodated three homogenous variables (estimated through factor loading) to form factor 5.This factor has contributed 5.64 per cent of variance and has been renamed as Educational capability.

Educational capability enhances the technology adaptation ability and hence improves the socialization of technology for higher income, better adaptability and long term sustainability.

Factor 6 has accommodated four functional variables (estimated through factor loading) to form factor 6.This factor has contributed 5.4 per cent of variance and has been renamed as farm capacity.

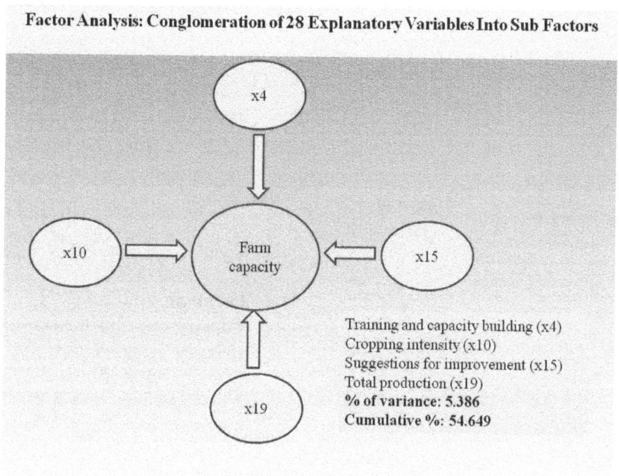

Model

Farm capacity endorses and enhances proper response and application of befitting technology towards improving productivity, income and livelihood.

Factor 7 has accommodated three homogenous variables (estimated through factor loading) to form factor 7.This factor has contributed 4.9 per cent of variance and has been renamed as Ecological proficiency.

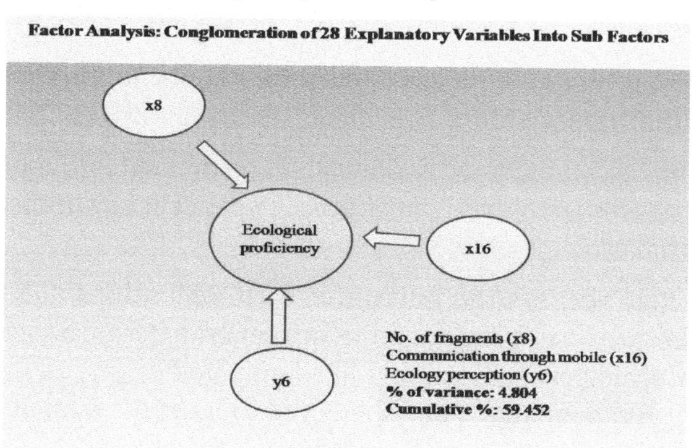

Model

Ecological services: production, maintenance, control, entertainment are all have been the subject to the unleashing of ecological proficiency during and after the production process organized at the farm.

Factor 8 has accommodated four functional variables (estimated through factor loading) to form factor 8.This factor has contributed 4.6 per cent of variance and has been renamed as Homestead dynamics.

Factor Analysis: Conglomeration of 28 Explanatory Variables Into Sub Factors

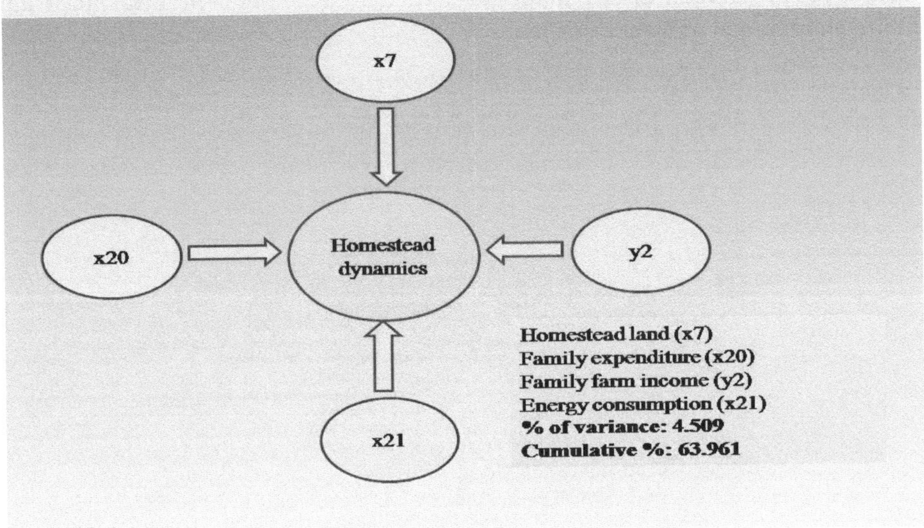

Model

Homestead land, mostly overlooked in India, could have been the most important functional support to both family and farm. That's why it's generated substantive contribution.

Factor 9 has accommodated one variable (estimated through factor loading) to form factor9.This factor has contributed 4.28 percent of variance.

This factor has made just linear and obvious contribution to the income and livelihood of the farmers.

Factor 10 has accommodated two variables (estimated through factor loading) to form factor 10.This factor has contributed 3.8 per cent of variance and has been renamed as social ecology.

Social ecology renders the fulcrum of farm operations and its productive behaviour of the farm itself. The decisions are generated from a social ecology that combine other ecological components like land, biodiversity to ultimately make ecological services adequately unleashed of to support the humans civilization.

Factor Analysis: Conglomeration of 28 Explanatory Variables Into Sub Factors

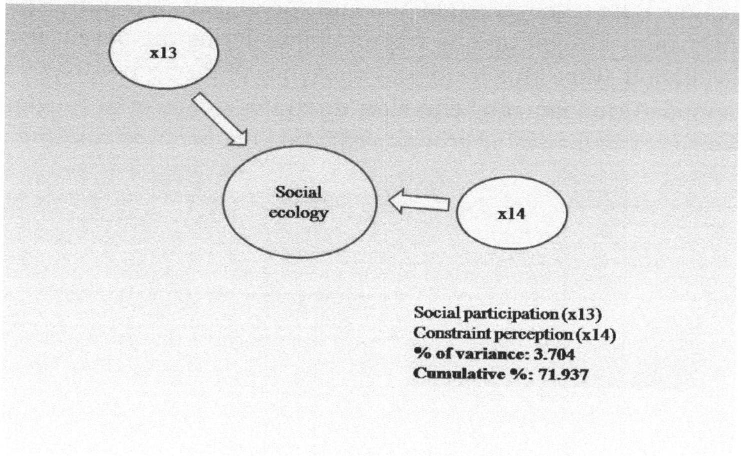

Model

* Variance has been calculated by dividing the Eigen value by the number of variables.

* Factor analysis assumes that the variance can have partitions into two parts: (i) common and (ii) unique.

* Common variance is the amount of variance that is shared among the set of items. A unique variance is shared by a solitary variable only.

* F1 to F10 except F9 are shared variance whether F9 has presented a unique variance as it is solitary variable.

Cluster Analysis

As derived from the cluster analysis by using single linkage the whole of the respondents here has been accommodated in 5 homogenous clusters.

- Cluster 1 has accommodated 100 respondents
- Cluster 2 has accommodated 1 respondent
- Cluster 3 has accommodated 1 respondent
- Cluster 4 has accommodated 2 respondents
- Cluster 5 has accommodated 1 respondent

The cluster has been formed because of the homogeneity of the respondents and distances between five clusters as measurable by Euclidean distance.

So, these five clusters can be considered as strategic respondents, group for taking any kind of strategic interventions like training, enterprise grouping, community mobilization programme.

Cluster analysis here has helped identity homogenous and operating groups in a population. Here it comprises of 105 enterprising farmers. It is expected that each of the groups has got unique way of manoeuvring the output features, i.e. Income, livelihood, wage and ecological resilience of the agro-ecosystems under study. This kind of grouping can help selecting strategic enterprise conglomerating, higher proficiency in managing groups, reduction of operating cost and risk.

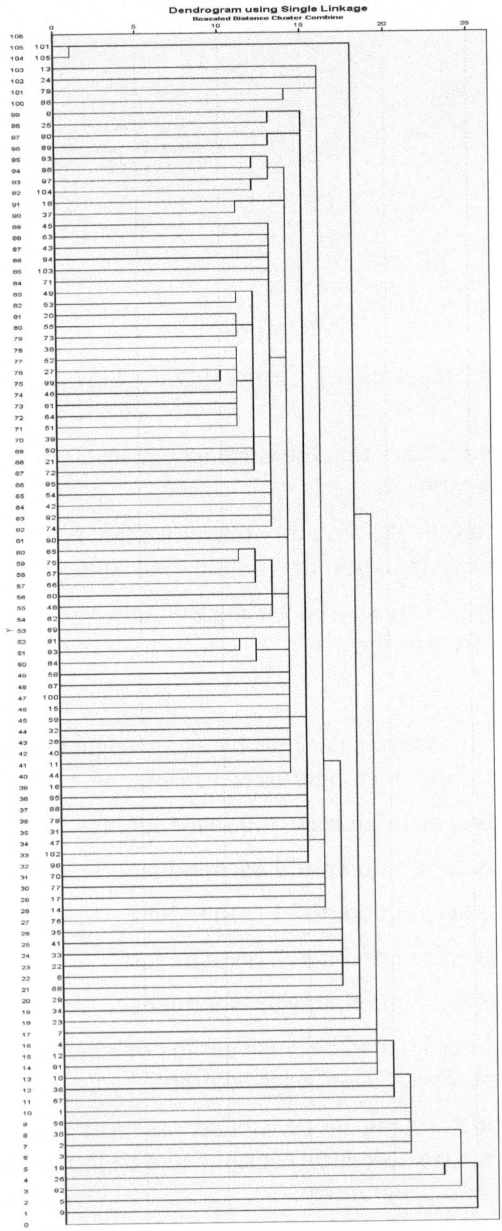

Artificial Neural Networking

Table: 28. Model Summary

Training	Sum of Squares Error		148.274
	Average Overall Relative Error		.696
	Relative Error for Scale Dependents	Average wage/Day (y_1)	.934
		Family farm income (y_2)	.704
		Productivity (y_3)	.803
		Income per bigha(y_4)	.687
		Man days (y_5)	.469
		Ecology perception (y_6)	.580
Testing	Sum of Squares Error		544.593
	Average Overall Relative Error		.984
	Relative Error for Scale Dependents	Average wage/Day (y_1)	1.005
		Family farm income (y_2)	.759
		Productivity (y_3)	.971
		Income per bigha(y_4)	.985
		Man days (y_5)	.318
		Ecology perception (y_6)	1.059

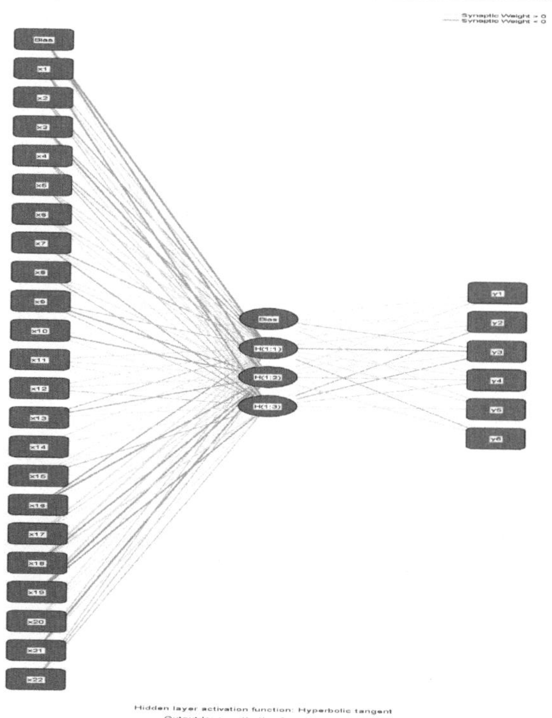

Artificial Neural Network

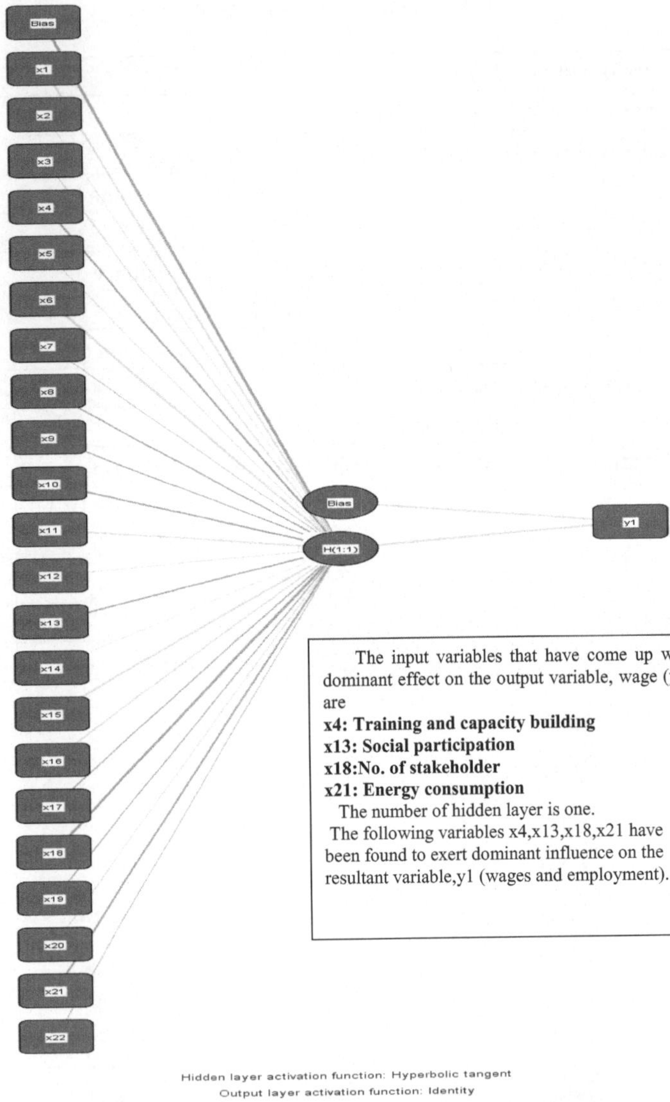

The input variables that have come up with dominant effect on the output variable, wage (y1) are

x4: Training and capacity building
x13: Social participation
x18:No. of stakeholder
x21: Energy consumption

The number of hidden layer is one.

The following variables x4,x13,x18,x21 have been found to exert dominant influence on the resultant variable,y1 (wages and employment).

Model Summary		
Training	Sum of Squares Error	3.088
	Relative Error	.087
Testing	Sum of Squares Error	407.035
	Relative Error	.836
Dependent Variable: Average wage/Day (y_1)		

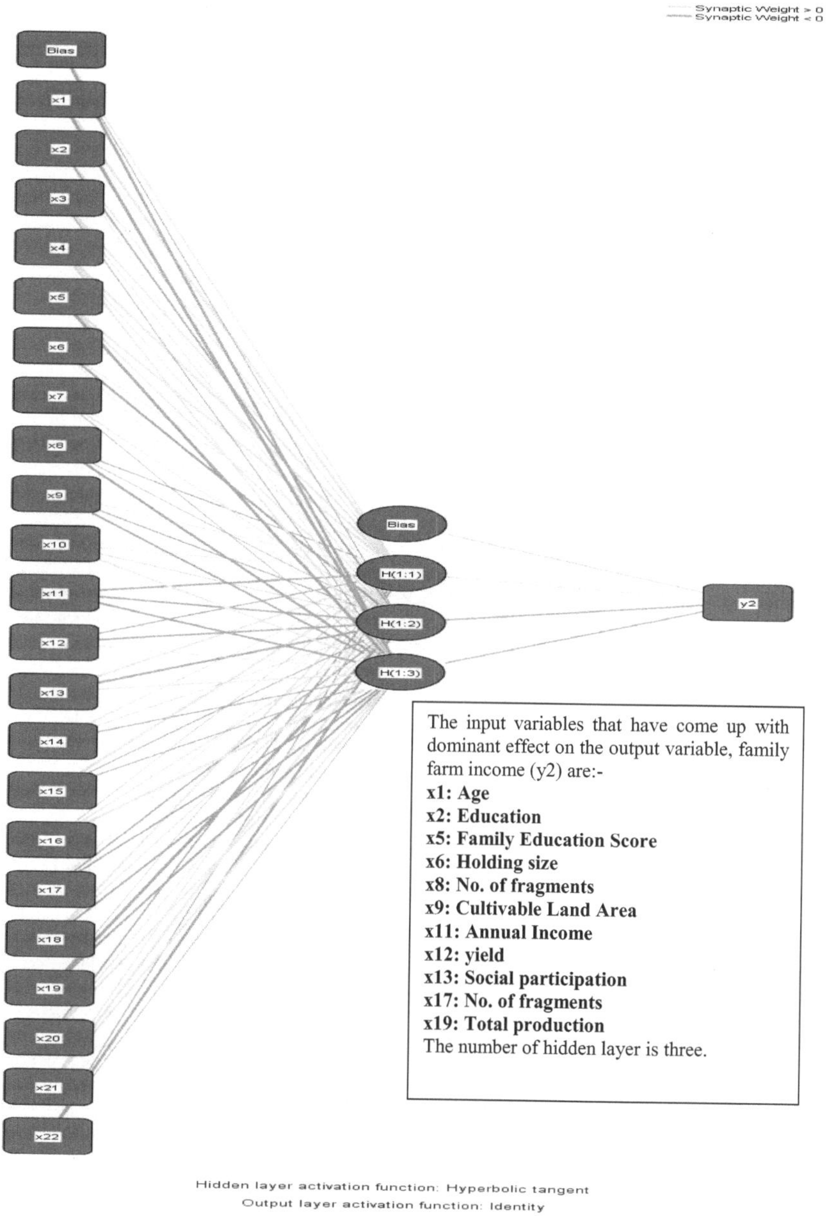

The input variables that have come up with dominant effect on the output variable, family farm income (y2) are:-

x1: Age
x2: Education
x5: Family Education Score
x6: Holding size
x8: No. of fragments
x9: Cultivable Land Area
x11: Annual Income
x12: yield
x13: Social participation
x17: No. of fragments
x19: Total production

The number of hidden layer is three.

Model Summary		
Training	Sum of Squares Error	29.140
	Relative Error	.767
Testing	Sum of Squares Error	5.491
	Relative Error	.629

Dependent Variable: Family farm income (y_2)

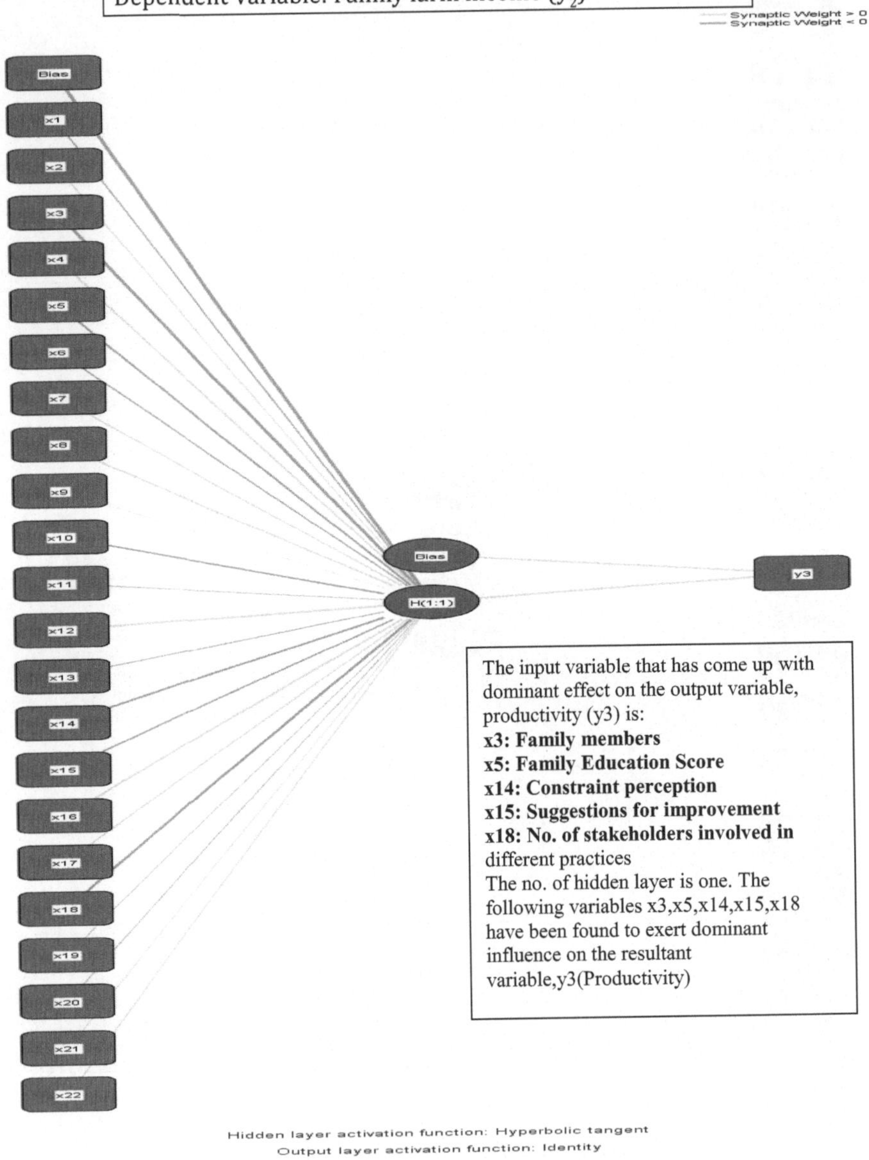

Model Summary		
Training	Sum of Squares Error	3.883
	Relative Error	.104
Testing	Sum of Squares Error	31.884
	Relative Error	.667

Dependent Variable: Productivity (y_3)

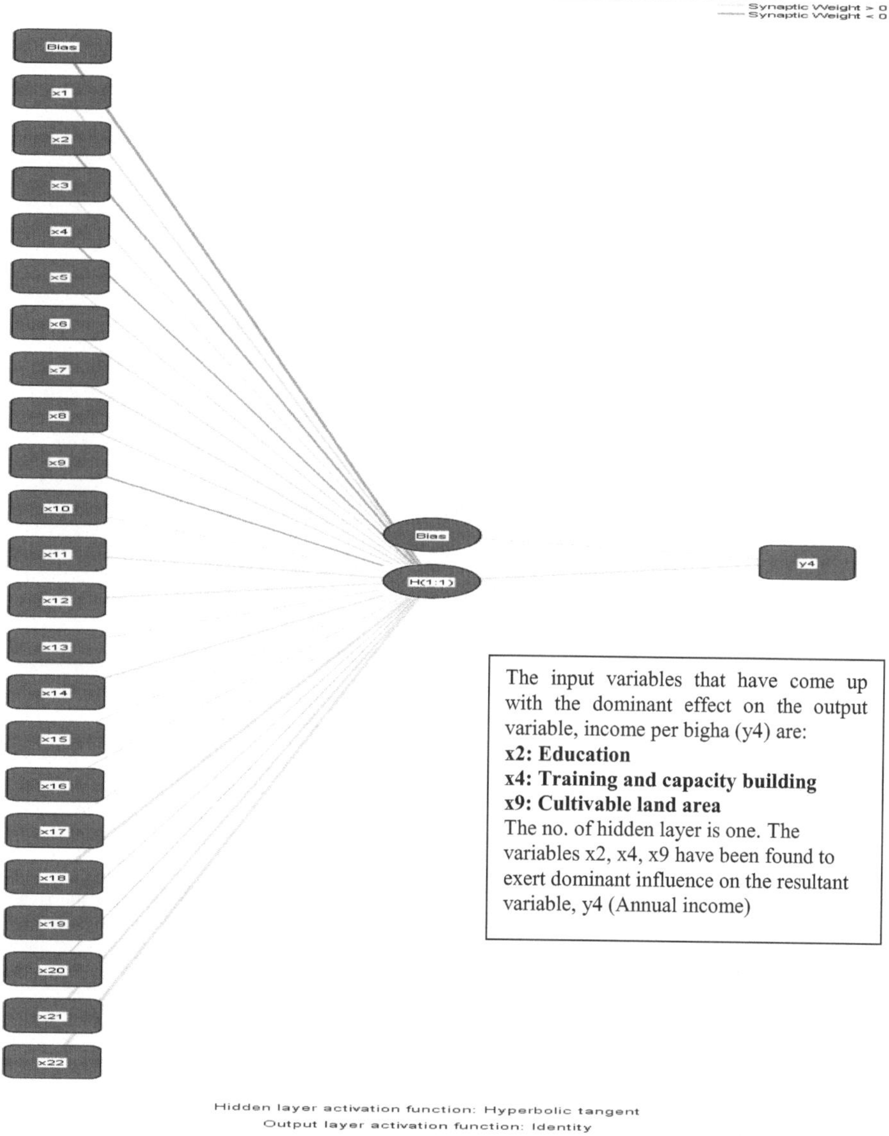

The input variables that have come up with the dominant effect on the output variable, income per bigha (y4) are:
x2: Education
x4: Training and capacity building
x9: Cultivable land area
The no. of hidden layer is one. The variables x2, x4, x9 have been found to exert dominant influence on the resultant variable, y4 (Annual income)

Model Summary		
Training	Sum of Squares Error	19.289
	Relative Error	.576
Testing	Sum of Squares Error	3.508
	Relative Error	.749

Dependent Variable: Income per bigha(y_4)

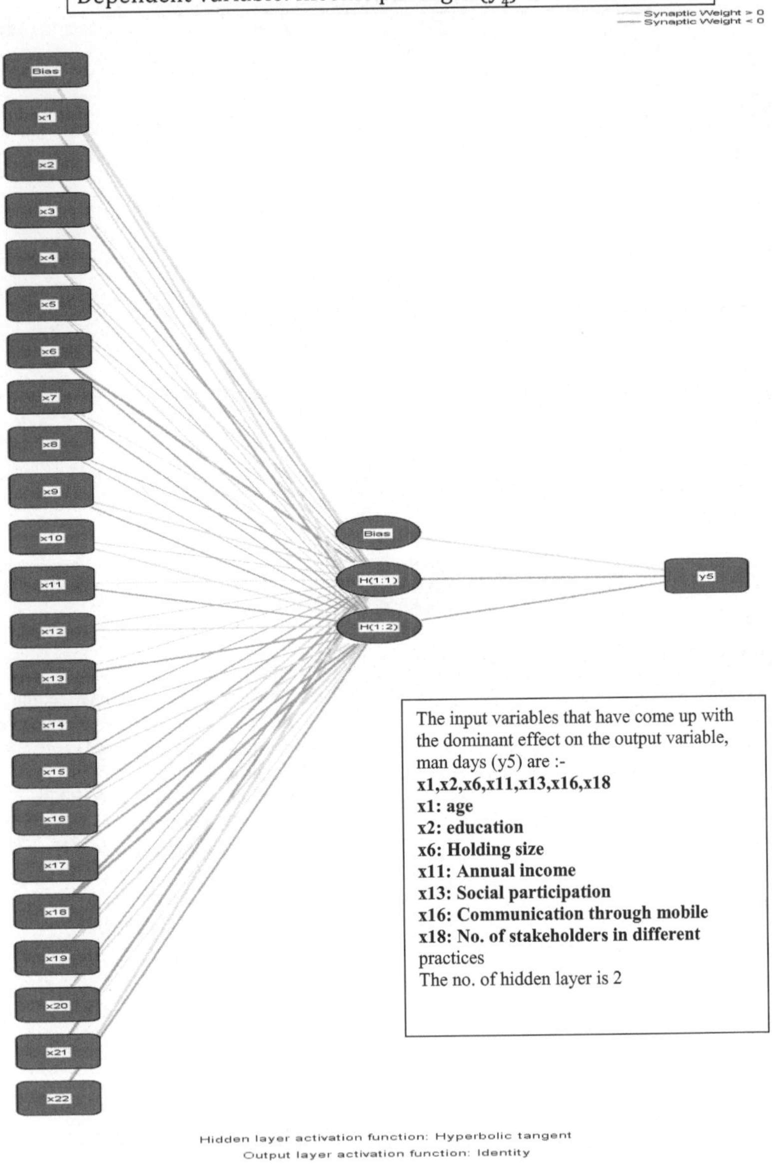

Model Summary		
Training	Sum of Squares Error	2..620
	Relative Error	.066
Testing	Sum of Squares Error	5.801
	Relative Error	.159
Dependent Variable: Ecology perception (y_5)		

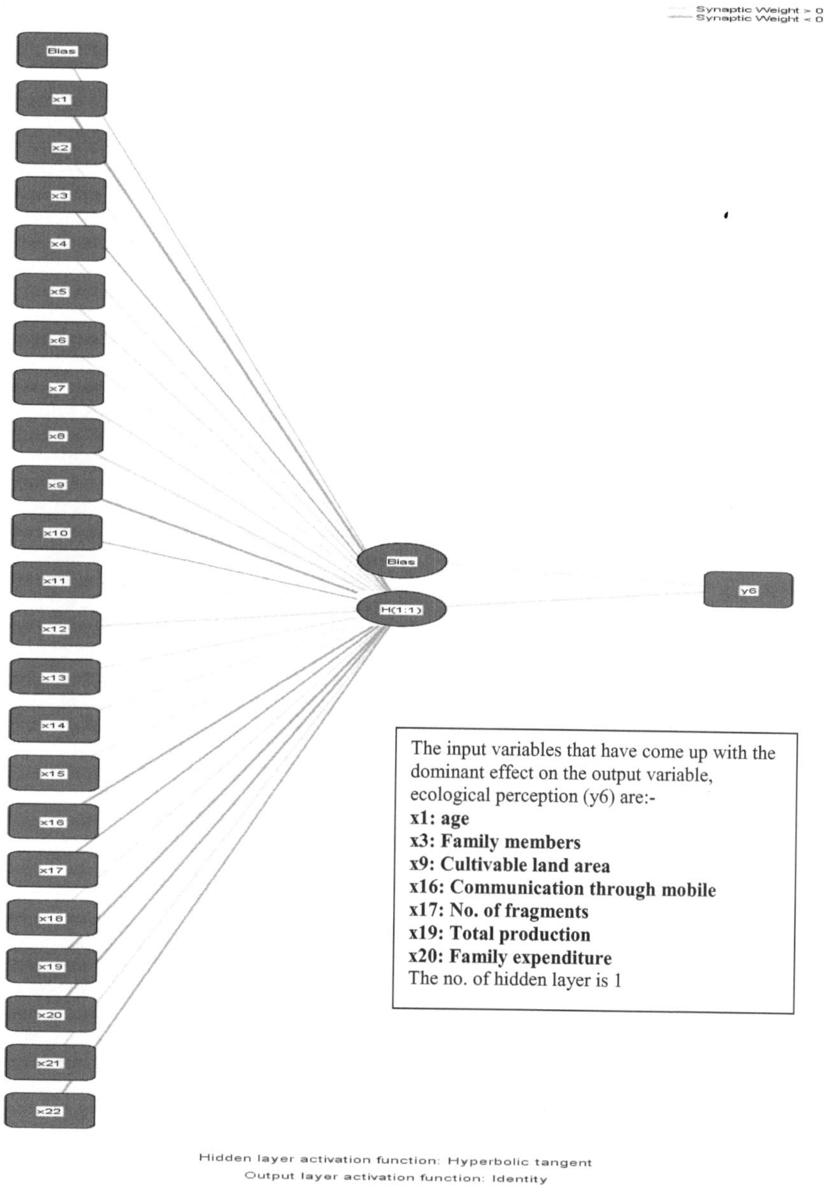

The input variables that have come up with the dominant effect on the output variable, ecological perception (y6) are:-

x1: age
x3: Family members
x9: Cultivable land area
x16: Communication through mobile
x17: No. of fragments
x19: Total production
x20: Family expenditure

The no. of hidden layer is 1

Model Summary		
Training	Sum of Squares Error	28.647
	Relative Error	.764
Testing	Sum of Squares Error	19.222
	Relative Error	.837
Dependent Variable: Ecology perception (y_6)		

The paradigm created out of stepwise Regression analysis with huge strategic implication.

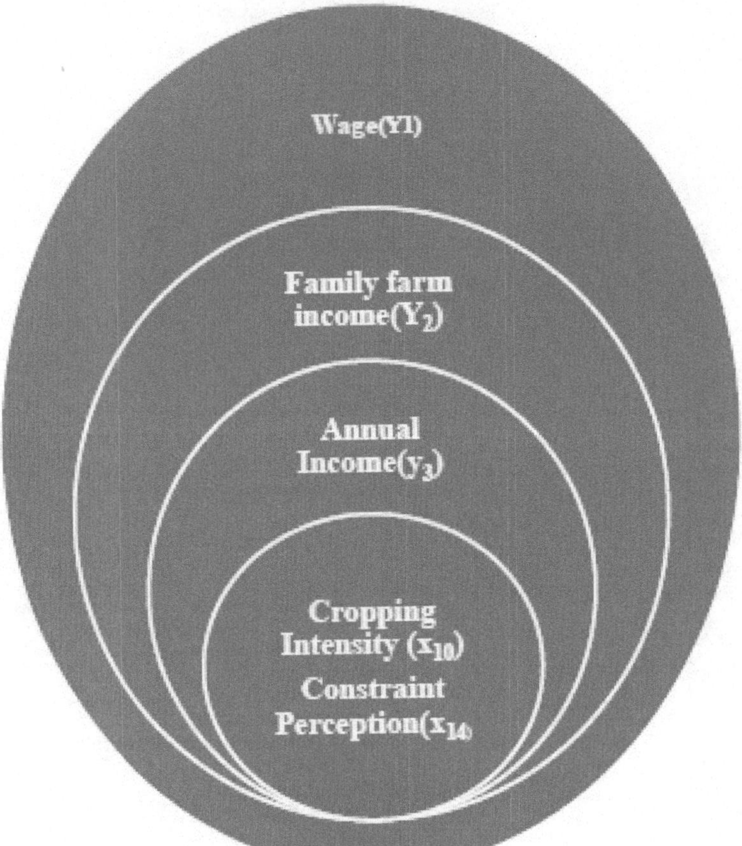

The common and the strongest determinants, as have been found in stepwise regression analysis,have been cropping intensity and constraints perception. So, the three predicted variables viz. Wage (y_1), family farm income (y_2) and income per bigha (y_4) can be simultaneously and isochronously dealt with cropping intensity and constraint perception. Or, in brief, income and wage both can be changed with higher cropping intensity and better management of constraints.

Chapter 7

SUMMARY AND CONCLUSION

Indian agriculture now stands at a cross road. By 2050, Indian population will overtake that of China's. The world population in the same year will be close to 9.5 billion. And to feed the human population of this planet, the agricultural production needs to be increased 60% over the present one by 2050. For India the target is 450 million tons by 2050.

The present study envisages as how to and whether the following factors *viz.* productivity, income, livelihood and ecological resilience can be optimally integrated on a mutually agreed upon platform of an operating agro-ecosystem. There is need to elucidate and estimate the kind, nature and impact of contributory factors on the very performance of ecological functions, so that productivity will generate livelihood and income. There should also be a policy implication as an expected outcome of the research to drive agricultural production, ecologically viable and economically productive. When this diode is being formed, the wage, income, and resilience are being maintained as an organic expansion of agricultural production, process and Management. The syndromes are very conspicuous in areas of highly depletive agriculture like Punjab, Haryana, Tamil Nadu, Andhra Pradesh and West Bengal as well. The indication of diminishing productivity,reducing livelihood and income redundancy are becoming prominent and alarming. So, a diverse farming system by replacing prototype agriculture is not only be agricultural reality but also an ecological necessity to sustain all kinds of micro, meso and mega life forms.

That is why and how the study of fishery delving into the need, necessity and niche of fish enterprises imbibing ecological, economic and social milieus has been a modern approach towards redeeming the value of farming system approach to be followed specially by the manager of miniscule land resources.

In the changed global scenario and with the dent for diversification of agricultural enterprises getting competitive in the global market, management of both quantity and quality are of prime issues in fish enterprise. The present study envisages assessing the productive performance of fresh water fisheries (fingerlings) in the defined domain of yield, income and employment generation and with a specific objective of taking the fish farmers as the respondents, who are managing it primarily at a domestic scale. While productive components of fish enterprise in terms of generation of yield, income and employment has been conceived as the consequent character, the causal character being a score of agro-economic, socio-personal and management by nature.

Objectives of the Study

- To delineate a general understanding of yield, income and livelihood of selected enterprises viz. Fish, rice and vegetables for the research locale.
- The interactions and impact of these three predicted characters viz. yield, income and livelihood in terms of a score of socio-economic and ecological predictors.
- To isolate the point of optimization or a common platform for landing wherein yield, income and livelihood will be behaving in a mutual incremental way to reach a threshold.
- To organize a series of participatory action learning for identifying the point of problem, solution, and planning.

Research Setting

Area of Study

The area of investigation of this study is situated in the state of West Bengal located in the eastern part of India. The state of West Bengal in eastern India has unique social and ecological backgrounds which influence the living standard and behavioral patterns of the people in many ways. The area of investigation belongs to the Haringhata block in Nadia district. The area of the study is ChakBirohi, Mitrapur, Bijra, Sekandarpur and Kastodanga of Nadia district.

Haringhata is a town in Kalyani subdivision of Nadia district in the state of west Bengal of India. Haringhata is located at 22°95′N 88°57′E. It has an average elevation of 10 meters. Haringhata community development block has an area of 168.59 Km^2.

Research Methodology

The present study has been conducted in one agro-ecological zones of the state West Bengal. Villages from Haringhata block in Nadia district and were selected from the new alluvial agro- ecological zone to conduct the study. The characters and the factors under study have been well discernible to this area

- The researcher's close familiarity with respect to area, people, officials and local dialects

- The ample opportunity to generate relevant data due to the close proximity of the area with the research and extension wing of the State Agricultural Universities;
- The highly cooperative and responsive respondents
- The profuse scope to get relevant information regarding perception on management of water resources, application of nutrients, improving soil organic carbon and problems and prospects of practicing conservation agricultural practices
- Experienced, well versed, venturesome, enthusiast and risk bearing farmers
- Scope of establishing a sustainable and profitable farming opportunities in the area

The study was conducted in the state of West Bengal. Haringhata block of Nadia district was selected for the study. In Haringhata block, a number of 5 villages was selected. For state, distrust and block selection, purposive sampling method was followed. A total of 105 respondents was intervened from the villages for the study purpose. Non-random snowball sampling was followed for respondents selection.

A pilot study has been conducted in the concerned area before constructing the data collecting devices. In course of this survey, informal discussion has been carried out with some farmers, local leaders, extension agencies and state agricultural officials of the localities. An outline of the socio-economic background of the farmers of the concerned villages, their interest in sustainable agricultural practices, experiences of sustainable farming and management strategies against the adversities helped in preparing and finalizing the interview schedule.

A set of 29 independent variables were taken for the study. Those are age of the respondent (x_1), education (x_2), family size (x_3), training and capacity building (x_4), family education score (x_5), holding size (x_6), homestead (x_7), no. of land fragments (x_8), cultivable land area (x_9), water bodies (x_{10}), weight of brooder fish (x_{11}), age at injection (x_{12}), scientific orientation (x_{13}), frequency of injection (x_{14}), input management and cost (x_{15}), other cost input management and cost(x_{16}), income statements (x_{17}), productivity (x_{18}), area under rice (x_{19}), area of vegetable (x_{20}), social participation (x_{21}), constraint perception(x_{22}), suggestions for improvement (x_{23}), communication through mobile (x_{24}), no. of fragments (x_{25}), no. of stakeholders (x_{26}), family expenditure (x_{27}), energy consumption (x_{28}), nutrition and health status (x_{29}). The dependent variables were average wage/day (y_1), family farm income (y_2), productivity (y_3), income (y_4), mandays (y_5), ecology perception (y_6).

The data collected in the interview schedule have been processed and analyzed in accordance with the outline laid down for the purpose at the time of developing the research plan. Processing implies editing, coding, classification and tabulation of collected data.

Various descriptive statistics viz., mean, standard deviation, coefficient of variation were used. The inferential statistics used were coefficient of correlation, multiple regression, stepwise regression, path analysis, canonical covariate analysis, factor analysis, principle component analysis, cluster analysis, artificial neural network analysis.

Salient Findings

Coefficient of correlation: Average wage/day (y_1) Vs. 22 Causal Variables (x_1-x_{22})

It is interesting to note that cropping intensity has recorded is significant net a negative correlation with wage and employment. This indicates that escalation of cropping intensity does not reflect on a positive change in wage and employment.

Constraint perception (x_{14}) has recorded a positive influence on wage and employment(y_1).It signifies that to a better understanding on the constraints confronted by him, he perhaps able to solve it and can successfully manage his farm with higher wage generation and employability.

Coefficient of correlation: (y_2) Vs. 22Independent Variables

Holding size(x6) has been found to exert significant and positive influence on family farm income.It is quite obvious that the bigger is the hodingsize,the higher has been the Family Farm Income.

Communication through mobile (x_{16}) and Energy consumption (x_{21}) however have recorded negative and significant influence on Family Farm Income while ecological level of energy consumption has contributed positive change in Family Farm Income,Mobile telephonic could not record any impressive effect on improving Farm Income.

On the other hand,no. of fragments (x_{17}) and no. of stakeholder in different practices (x_{18}) have positively influenced.To improve the Family Farm Income(x_2) small fragment of land here have been found to have higher proficiency in generalizing Family Farm Income and the more number of stakeholder involved in farm production process have also exerted positive influence on Family Farm Income.

Coefficient of Correlation (r): (y_3)Vs. 22Independent Variables

Training and Capacity Building (x_4) has recorded positive and significant correlation. This indicates that intensive training has been instrumental in improving the productivity of farm. Yield of crops (x_{12}) has collectively contributed to the overall productivity of farm and the relation is obvious. No. of stakeholders (x_{18}) evinced that for leisure no. of stakeholder,the productivity has gone up.

Coefficient of Correlation (r): (y_4)Vs. 22Independent Variables

Family with higher number of members are engaged in different types of agricultural operations. It is also found that holding size, homestead land, no. of fragments have generated more annual income. It is interesting to note that the higher the family income, the more is the annual income. The farms generating higher annual income are also engaging more number of stakeholders having better nutrition and health status. So, the farm production goes better.

Coefficient of Correlation (r): (y_5) vs. 22 Independent Variables

It has been found that farmers with higher age echelon have generated more mandays than the younger ones. The other side of the fact is that young generation rural people are increasingly withdrawing themselves from agriculture. It is elicited that those who are generating more man days they are prone to interpersonal communication than to access mobile telephony. The farms generating higher man days are also engaging more and diverse stakeholders having higher nutrition and f better health status and in the end,their farm production goes better.

Coefficient of Correlation (r): (y_6)Vs. 22Independent Variables

It has been found that communication through mobile has recorded significant and negative correlation with ecology perception.

Mobile communication,byit's nature can impact has recorded a negative correlation with ecology perception. Prolonged and unscientific use of mobile telephone has d got an adverse effect on health and surrounding ecosystem as well.

Stepwise Regression Analysis: (y_1) Vs. 22 Causal Variables

It has been found that cropping intensity has recorded significant and negative correlation with wages per day.

Constraint perception (x_{124}) has recorded significant and positive correlation with wages per day.

Stepwise Regression Analysis: (y_2) Vs. 22 Causal Variables

The table implies that cropping intensity (x_{10}) has recorded significant and negative correlation with family farm income.

Constraint perception (x_{14}) has recorded significant and positive correlation with income.

Stepwise Regression Analysis: (y_3) Vs. 22 Causal Variables

The following variables-Holding size has been retained as the most significant and highest contribution to the consequent character-Productivity.

This interestingly indicates that holding size is still the highest contributor in improving the productivity of farms. It is also discernible that land resource is at least for this case, has been the most important driver of productivity.

Stepwise Regression Analysis: (y_4) Vs. 22 Causal Variables

The stepwise regression analysis where two variables –cropping intensity (x_{10}) and constraint perception (x_{14}) has been retained at the last step.

Cropping intensity helps increase annual income substantively. When constraints are perceived properly and dealt with effectively, it helps increase income.

Stepwise Regression Analysis: (y_5) Vs. 22 Causal Variables

It has been found that two variables:-no. of stakeholders involved in different practices (x_{18}) and annual income (x_{11}) have been retained at the last step in the process of backward elimination.

It indicates that man days have been generated the most wherein no. of stakeholders (x_{18}) are on a higher side. With the diversification of farming through inclusion of elite enterprises like fishery, poultry, horticulture alongside the main field crop enterprise, the no. of man days has been increased.

Farmers with higher annual income have been found to generate higher intensity of man days.

Involvement of diverse stakeholders, crop diversification and higher size of holding have been found to generate a combined and positive effect on generation of both high intensity of man days and productivity as well.

Stepwise Regression Analysis: (y_6) Vs. 22 Causal Variables

It has been found that the following variables –communication through mobile,yield,cultivable land area, no. of fragments,total production have been retained at the last step.

Yield (x_{12}) is an ecological function so it has got a substantive contribution perception. Cultivable land (x_9) has a natural resource is well integrated to ecology. The nature and approach to management of the land has got correlates to energy management,methane emission,water economy and biodiversity conservation.

Path Analysis: Decomposition of Total Effect into Direct, Indirect and Residual Effect: Y_1

It has been found that income is the most important determinant for a farmer from his farm as to whether his income will go after generations the wage and employment with higher income he will be able to hire more labours and can ensure better input and management.

The highest indirect effect has been exerted by the variable no. Of stakeholders involved in different practices. This implies that there has been an orchestration of multi-stakeholder participation making farm economics successfully generating higher income with a simultaneous creation of wage and employment.

Path Analysis: Decomposition of Total Effect into Direct, Indirect and Residual Effect: Y_2

Holding size has been recorded the highest substantive effect on farm income. So,it is clear that the holding size as resource endowment owned by the farmer has become the strong determinant to improve the farm income. It is further to infer that even today farm income is very much scale sensitive and goes better for the bigger size. The indirect effect,however,has been executed by the no. of stakeholders. The connectivity and interaction by a farmer with multistakeholder entities have contributed positively to make farmer successful enterprise.

The variable Holding size (x_6) has routed the highest indirect effect of as many as 13 exogenous variable to ultimately characterized the behaviour and performance the consequent variable i.e. Family Farm Income.

Path Analysis: Decomposition of Total Effect into Direct, Indirect and Residual Effect: Y_3

The variable yield of crops (x_{12}) has contributed the highest direct effect on the productivity, and this is for an obvious reason that when yield of individual crops increases, the productivity, of the farm as a whole improves.

The annual income as exerted the highest indirect effect on productivity. This may derive an element of doubt that whether the income from farm has proportionately been recycled into the overall process of farm entrepreneurship.

The variable-Holding size (x_6) has routed the highest indirect effect of as many as 10 causal variables to ultimately characterize the performance of the consequent variable, productivity. So in most cases, whatever may be the contributory factors, it needs to accompany the associate property of holding size.

Path Analysis: Decomposition of Total Effect into Direct, Indirect and Residual Effect: Y_4

The variable-Family members (x_3) has exerted the highest indirect effect. It means that the size of the family is an important determinant exertive intensive associating property to settle the family income.

On a simpler note,it is to infer that the farm income has been impacted by both potential consumer and potential contributor in the form of family labour.

The variable annual income (x_{11}) has got a direct and substantive effect impacts on on farm income. The higher is the family income,the the better and stronger are the investment ability which results into higher farm inome.

An assure and higher family income ensures adequate and proper time of application to exel the proficiency of input management which reflects in the farm of higher yield and better income.

It implies that combination of 22 exogenous variable could not explain 26.59 per cent of variance in the consequent variable, **income(y4).**

Path Analysis: Decomposition of Total Effect into Direct, Indirect and Residual Effect: Y_5

It has been found that no. of stakeholders involved in different practices (x_{18}) has exerted the highest direct effect and annual income, on the other hand, has exerted the highest indirect effect. Again, the variable- no. of stakeholders involved in different practices has exerted the highest associative impact.

It has been found that the variable, no. of stakeholders (x_{18}), has enrooted on man days (y_5) the highest indirect effect on as many as 16 exogenous variables to ultimately impact on the consequent variable man days (y_5).

It implies that the combination of these 2 exogenous variables could not explain 14.64 percent of variance in the consequent variable, mandays (y_5).

Path Analysis: Decomposition of Total Effect into Direct, Indirect and Residual Effect: Y_6

No. of fragments (x_{17}) has contributed the substantive impact of ecological perception. Since, unabated fragmentation keeps on contributing to energy and resource prodigal nature of holdings and thereby generating a deleterious impact on ecological resilience, it has generated a strong impact on ecology perception.

Canonical Covariate Analysis: Interaction between Left Side Dependent and Right Side Independent Variables

The following conglomeration of y variables (y_1), (y_2), (y_3), (y_6) are moving together and these conglomeration of variables have selectively been interacting with the following sets of variables (x_1, x_2, x_3, x_4, x_5, x_6, x_7, x_8, x_9, x_{10}, x_{11}, x_{12}, x_{13}, x_{14}, x_{15}, x_{16}, x_{17}, x_{18}, x_{19}, x_{20}, x_{21}, x_{22}). So, an isochronous change in wage (y1), income (y2), productivity (y3) and ecology perception (y6) merits a concurrent contribution from the following exogenous variables.

Similarly, the other two dependent variables –farm income and man days are moving together. The lion's ear of cost in farming is incurred in the form of man days. So, income from farm and man days are moving together and this relation and interaction –a concurrent contribution is coming from the following x variables.

The intra mode of interaction is based on cross loading. Here, in this Canonical Covariate Analysis, the y-set of variables has been regressed in terms of x-set of variables and on the contrary, the x-set of variables has been regressed in terms of variables.

Factor Analysis: Strategic Conglomeration of Variables into Factor

Factor 1 has accommodated four homogenous variables (estimated through factor loading) to form factor1.This factor has contributed 20.05 per cent of variance and has been renamed as farm economy.

Factor 2 has accommodated three homogenous variables (estimated through factor loading) to form factor 2.This factor has contributed 9.09 per cent of variance and has been renamed as resource character. The complex and non-linear interactions amounts in between resource variables under the factors have contributed to the income level, income, productivity and management to the entire production regime and it's response to market.

Factor 3 has accommodated two homogenous variables (estimated through factor loading) to form factor 3.This factor has contributed 8.12 per cent of variance and has been renamed as productive function.

Productive function is the prime to make a system and their interaction, successfully performing and that is how this variable has contributed significantly to the total variance.

Factor 4 has accommodated two homogenous variables (estimated through factor loading) to form factor 4.This factor has contributed 6.40 per cent of variance and has been renamed as family health.

Health and nutrition of farmers, engaged in innumerable farm operations, supports the productive performance of a farm, small, big or medium.

Factor 5 has accommodated three homogenous variables (estimated through factor loading) to form factor 5.This factor has contributed 5.64 per cent of variance and has been renamed as Educational capability.

Educational capability enhances the technology adaptation ability and hence improves the socialization of technology for higher income, better adaptability and long term sustainability.

Factor 6 has accommodated four homogenous variables (estimated through factor loading) to form factor 6.This factor has contributed 5.4 per cent of variance and has been renamed as farm capacity.

Farm capacity endorses and enhances proper response and application of befitting technology towards improving productivity, income and livelihood.

Factor 7 has accommodated three homogenous variable (estimated through factor loading) to form factor 7.This factor has contributed 4.9 per cent of variance and has been renamed as Ecological proficiency.

Ecological services: production, maintenance, control, entertainment are all have been the subject to the unleashing of ecological proficiency during and after the production process organized at the farm.

Factor 8 has accommodated four homogenous variables (estimated through factor loading) to form factor 8.This factor has contributed 4.6 per cent of variance and has been renamed as Homestead dynamics.

Factor 9 has accommodated one variable (estimated through factor loading) to form factor9.This factor has contributed 4.28 percent of variance.

This factor has made just linear and obvious contribution to the income and livelihood of the farmers.

Factor 10 has accommodated two variables (estimated through factor loading) to form factor 10.This factor has contributed 3.8 per cent of variance and has been renamed as social ecology.

- * Variance has been calculated by dividing the Eigen value by the number of variables.

- * Factor analysis assumes that the variance can be partitions into two parts: (i) common and (ii) unique.
- * Common variance is the amount of variance that is shared among the set of iems. A unique variance is shared by a solitary variable only.
- * F_1 to F_{10} except F_9 are shared variance whether F9 has presented a unique variance as it is solitary variable.

Cluster Analysis

As derived from the cluster analysis by using single linkage the whole of the respondents here has been accommodated in 5 homogenous clusters.

The cluster has been formed because of the homogeneity of the respondents and distances between five clusters as measurable by Euclidean distance.

So, these 5 clusters can be considered as a strategic respondent group for taking any kind of strategic interventions like training, enterprise grouping, community mobilization programme.

Artificial Neural Networking

The common and strongest determinants, as have been found in stepwise regression, are cropping intensity and constraints perception. So, the three predicted variables viz. Wage (y_1), family farm income (y_2) and Annual Income (y_4) can be simultaneously dealt with cropping intensity and constraint perception. Or, in brief, income and wage both can be changed with higher cropping intensity and better management of constraints.

Future scope of the study

The entire empirical study has incubated and instigated to go for further research in following domains with relevance of replication.

- In the future study, new enterprises, echelons and ecological niche are proposed to include.
- Simulation based modeling based on both history and training data can be another area of research on income, productivity, ecological resilience and sustainable livelihood.
- The process of resource, function, cost and management optimization can well invite and extrapolation of present research.
- A well guided policy research on orchestration of productivity, livelihood and ecology can go a long way in shaping the destiny of millions of resource poor farmers in Indian beyond.

Limitations

- More no. of relevant variables could have been included.
- It needs more quality research along with quantitative research.

- Participatory research and data generation are missing somehow.
- Application of deep learning in addition to Artificial Neural Network (ANN) could have been helpful in eliciting and designing quality interpretation and reliable prediction.
- The no of respondents could have been escalated but due to pandemics it has been mostly non-random made of selections and inevitable the no. of respondents has been restricted.

Conclusion

The sociology, ecology, economy and productivity mostly moving mutual contradiction. It is extremely difficult to get them operating in perfect synchronization and reconciliation. The burning issue in Indian agriculture lies into it's inherent conflict between productivity and ecology, income and productivity, livelihood and productivity and also productivity versus distributive justice.

There have been so many examples in Indian agriculture wherein good harvest amounts to bad return, or a successful harvest has contributed to fair amount of ecological disruption. This is a million dollar question as to how and whether the blasphemy of contradiction between productivity and ecology or ecology and livelihood can be transformed into a symphony of sociology-economy-ecology-livelihood.

The present study-"Yield, income and livelioood in selected farm enterprises: Estimation of socio-ecological compliances and conflicts." Has been carried out in five selected villages under Haringhata block (Latitude: 22°57'37.84"N; Longitude: 88°34'2.65"E) of Nadia district, West Bengal. The villages vis-à-vis research locale have been unique by faster modernization in agriculture with all possible elements of green revolution. The farmers are and have been well responsive to innovations and experimentation both in self-learning and inductive process of technology, socialization. A score of 105 farmers have been selected through both random and non-random sampling method. Two are mainly small and marginal farmers venturing through innovations , experimentation and entrepreurships.

The empirical study depicts that the unique conciliation and conflict between and amongst sets of exogenous and results and variables. The variables- size of holding, no. of stakeholders, annual income, health, and nutrition, cropping intensity, perception on constraints, energy consumption have been found to exert strong and decisive impact on different planks of social ecology of productivity in agriculture. The marker variables have characterized the nature, intensity and direction of binary interactions between productivity, income, livelihood with a set of 22 exogenous variables. The following factors- farm economy, resource character, productive function, family health, educational capability, farm capacity, ecological proficiency, homesteads dynamics, social ecology as extracted, have got tremendous strategic importance for delineating micro level policy implication and intervention as well.

The sociology of agriculture remains in seamless hunch for exploring a common valley for landing wherein ecology, economy, sociology will ensure a concentrated integration and symphony. It will, as we hope, for each golden harvest, there must be a happy income to bring a decent life and livelihood for each of the farmers in India.

Glimpses of Field Interaction for Data Collection and Visualization

Field data collection on brinjal enterprises at village – Ghoragacha. Market price, productivity and disease pest problem of brinjal have been the prime concern (2019).

Data collection on boro rice and interaction with farmers. The crux of the discussion has been productivity and market price of boro paddy. The issues of man days, water management and marketability have also been documented (village-Chak Birohi, 2021).

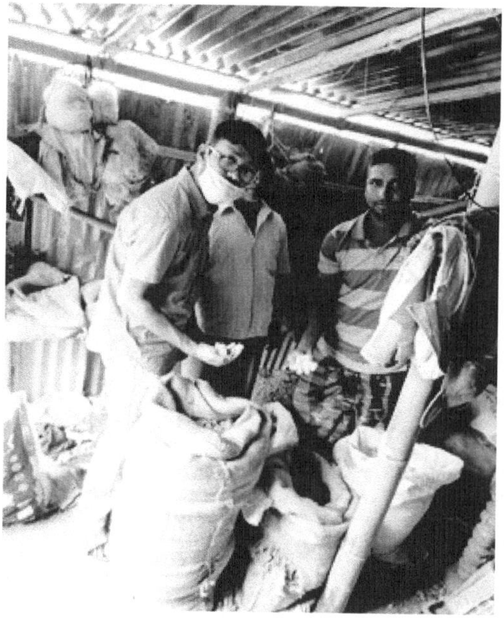

Data generation on fish enterprises in terms of income, productivity, wage, livelihood and ecology. Availability of fish feed and fingerlings have been prime issues (Village- Mitrapur, 2020).

Data collection on Cabbage enterprise at Kantabelia village in regard to production, marketability, disease pest problem, adverse weather condition and livelihood (2021).

Livelihood in Agriculture lays the foundation of employment for rural India. Here in West Bengal, more than 90 per cent of the farms are marginal and fragmented too. The field data collection covers and delineates the opportunity of livelihood generation for both male and female (Village: Kantabelia, 2020)

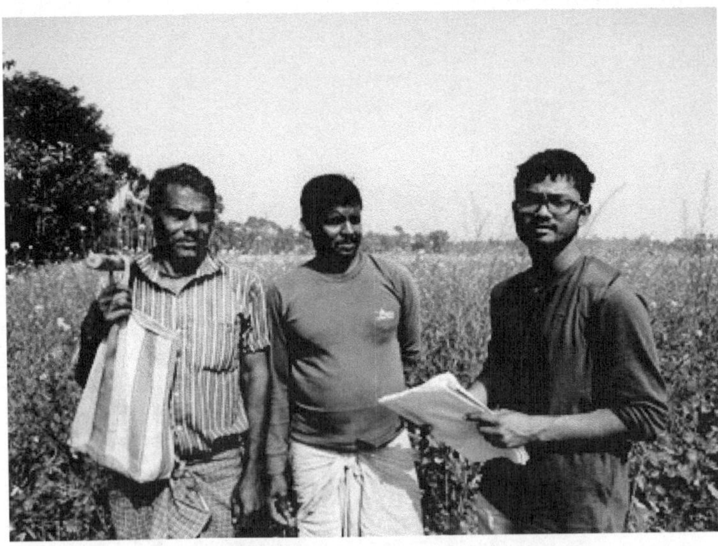

Mustard and pulses are serendipitously replacing summer rice to economize cost and risk and at the same time to save groundwater and ecological resilience. The data generation covers productivity, profitability and livelihood associated with mustard. The ecological dividends have also been the crux of discussion. (Village: Ghoragacha, 2021)

Livestock plays the pivotal for agrarian economy of India. It provides income, nutrition, livelihood, ecological services and what not! The study village is maintaining yet a good count of livestock population which provides fillip to the economy of the village. The data collected on livestock covers a kaleidoscopic vision on multidimensional contribution including income, livelihood and ecology (Village: Ghoragacha, 2021).

Data collection from villagers cannot be insipid and instrumental. It needs human face and science of communication. While in the fishery field, data have been collected on income, productivity, ecology and marketability of fish enterprises from a successful fish entrepreneur (Village: Mitrapur, 2020)

Chapter-8

FUTURE SCOPE OF THE STUDY

The Entire empirical study has incubated and instigated to go for further research in following domains with relevance of replication.

- In the future study, new enterprises, echelons and ecological niche are proposed to include.
- Simulation based modeling based on both history and training data can be another area of research on income, productivity, ecological resilience and sustainable livelihood.
- The process of resource, function, cost and management optimization can well invite and extrapolation of present research.
- A well guided policy research on orchestration of productivity, livelihood and ecology can go a long way in shaping the destiny of millions of resource poor farmers in Indian beyond.

Limitations

- More no. of relevant variables could have been included.
- It needs more quality research along with quantitative research.
- Participatory research and data generation are missing somehow.
- Application of deep learning in addition to Artificial Neural Network (ANN) could have been helpful in eliciting and designing quality interpretation and reliable prediction.
- The no of respondents could have been escalated but due to pandemics it has been mostly non-random made of selections and inevitable the no. of respondents has been restricted.

- Since the R^2 values have been found low in stepwise regression analyses, it has been essential to go after CCA and ANN analyses to understand the nature of direction of interactions.

BIBLIOGRAPHY

Acharya, S. S. (1980). *Agricultural Marketing In India,* (Ed).Higher Productivity in Agricultural,Oxford and B.H.PublishingCompany,New Delhi.

Acharya, S. K., Pal, P. and Biswas, A. (2020). Research methodology: the design, process and application. Satish Serial Publishing House, Delhi.

Acharya, S.S., (2006) Sustainable Agriculture and Rural Livelihoods, Monograph No. 6, ICSSR Occasional Monograph Series, Indian Council of Social Science Research, New Delhi.

Ahnstrom,J., J. Hockert,H.L. Bergea,C.A. Francis, P. Skelton et al. 2009. Farmers and nature conservation: what is known about attitudes, context factors and actions affecting conservation? Renewable Agriculture and Food Systems 24(1):38 47.Cambridge, UK: Cambridge University Press.

Alderman, Harold.(2007. —Managing Risk to Increase Efficiency and Reduce Poverty.|| Background Paper for the World Development Report 2008.

Allison, E.H., B.D. Ratner, B. Asgard, R. Willmann, R. Pomeroy et al. 2012. Rights based fisheries governance: from fishing rights to human rights. Fish and Fisheries 13(1): 14 29.Oxford, UK: Wiley Blackwell.

Anderson, Kym, editor. 2009. Distortions to Agricultural Incentives. A Global Perspective, 1955 – 2007. New York and Washington D.C.: Palgrave Macmillan and World Bank.

Anonymous, (1992), Management of heliothis on cotton in Andhra Pradesh. All IndiaCoordinated Cotton Improvement Project, Silver Jubilee Sym., Ludhiana, 1992.

Baffes, John, and Bruce Gardner. 2003. —The Transmission of World Commodity Prices to Domestic Markets under Policy Reforms in Developing Countries. Journal of Economic Policy Reform 6 (3): 159 -180.

Baffes, John, and TassosHaniotis. 2010. —Placing the2006/08 Commodity Price Boom into Perspective.|| Policy Research Working Paper No. 5371, World Bank, Washington D.C.

Bale, Malcolm and Ernst Lutz. 1979. The Effects of Trade Intervention on International Price Instability.|| American Journal of Agricultural Economics 61 (3): 512 – 516.

Balkrishna R. (1958) Measurement of Productivity in Indian Industry;UniversityPress,Madras

Balla S.S. (1979) Form Size, Productivity and Technical change in Indian Agriculture in Berry and Cline

Barrett Christopher B., Thomas Reardon and Patrick Webb. (2001). —Nonfarm Income Diversification and Household Livelihood Strategies in Rural Africa: Concepts, Dynamics, and Policy Implications.|| *Food Policy 26*: 3151 – 331.

Barrett, Christopher B. 2002. —Food Security and Food Assistance Programs.|| In Handbook of Agricultural Economics, Vol. 1B, ed. Bruce L. Gardner and Gordon C. Rausser. Amsterdam: North Holland.

Barrett, Christopher, Michael R. Carter and C. Peter Timmer. 2010. —A Century-Long Perspective on Agricultural Development.|| *American Journal of Agricultural Economics* 92 (2): 447 – 468.

Barun K. Pal (2019). *The Profitability of Paddy cultivation and Farmers 'Perception in West Bengal with Special Reference to PurbaBardhamanDistrict.*https: / / doi.org/ 10.20896/ saci. V7i3.537

Berry RA. & Cline W.R.(1979) Agrarian Structure and Productivity in Developing Countries;John Hopkins University Press,Baltimore.

Bharadwaj K. (1974)Production Conditions in Indian Agriculture;Cambridge University Press,Cambridge.

Brahmanda P.R. (1982) Productivity in the Indian Economy-Rising inputs for falling outputs;

Bravo Ureta, B.E., Solis, D., Cocchi, H., and Quiroga, R.E. (2006) The impact of soil conservation and output diversification on farm income in Central American hillside farming. *Agricultural Economics*, 35(3): 267-276.

Bravo-Ortega, Claudio and Daniel Lederman.(2005).||Agriculture and National Welfare around the World: Causality and International Heterogeneity since 1960. || Policy Research Working Paper No. 3499, World Bank, Washington D.C.

Breman, J. (2016). At work in the informal economy of India: A perspective from the bottom up (OIP). OUP Catalogue.

Burk M.C. & Ezekiel Maredeca (1967): Food and Nutrition in Developing Economy (In Agricultural Development & Economic Growth);Edited by Southern and Johnston,New York Camell University Press,P.337.

Byerlee, D., Harington, L. and Winkelmann, D.L., (1982), Farming systems research -Issues in research strategy and technology design. American J. Agric. Econ., 64(5):897- 904

Byerlee, Derek, Alain de Janvry and Elisabeth Sadoulet (2010). —Agriculture for Development – Revisited. Lessons learned from the University of California at Berkeley Conference.|| Paper presented at the Conference *Agriculture for Development-Revisited,* University o California at Berkeley, October 1-2

Byerlee, Derek, Alain de Janvry and Elisabeth Sadoulet. 2009. —Agriculture for Development: Background Note April 2008, Washington D.C.

Chatterjee, S., & Gupta, S. (2013). Extent of technological change in rice cultivation over four decades in West Bengal, India. *Agricultural Economics,* 59(6), 281-291.

Chowdhury, M. & Muvva, L. (2018). Impact of Farm Mechanization On Income And Employment and Constraints In Mechanization Of Rice Cultivation In West Godavari District. ANGRAU, 10, 107.

Coady, David, Margaret E. Grosh and John Hoddinott. (2003). *Targeting of Transfers in Developing Countries: Review of Lessons and Experience.* Washington D.C.: World Bank

Datt, Gaurav and Martin Ravallion. 1996. —How Important to India's Poor is the Sectoral Composition of Economic Growth? The World Bank Economic Review 10 (1): 1 – 25.

Datt, Gaurav and Martin Ravallion. 1998. —Farm Productivity and Rural Poverty in India.|| Journal of Development Studies 34 (4): 62 – 85.

De Janvry and Jean-Jacques Dethier. (1985.) Technological Innovation in Agriculture: The Political Economy of its Rate and Bias.: CGIAR Study Paper No. 1, World Bank,

de Janvry, Alain and Elisabeth Sadoulet (2009). Agricultural Growth and Poverty Reduction: Additional Evidence.|| *The World Bank Research Observer* 25 (1): 1 – 20.

Deasy, C., J.N. Quinton,M.Silgram,A.P.Bailey,B.Jackson, and C.J.Stevens.(2010). Contributing understanding of mitigation options for phosphorus and sediment to a review of the efficacy of contemporarayagricultural stewardship measures.Agricultural systems 103(2):105 109.Oxford,UK:Elsevier Ltd.

Demeke, Mulat, GuendalinaPangrazio and MaterneMaetz. 2008. Country Responses to the Food Security Crisis: Nature and Preliminary Implications of the Policies Pursued.|| Agricultural Policy Support Service, FAO, Rome.

Desai, B.K., Satyanarayan Rao, Biradar, S.A., Prahala, U., Shashikumar, M., Santosh,U.N. and Jgannath. (2013), Development of profitable integrated farming systemsfor small and marginal farmers of Hyderabad Karnataka region under irrigationcondition. Int.J.Agric.Environment and Biotechnology, 6(4): 617-622.

Dethier, Jean-Jacques & Effenberger, Alexandra. (2011). Agriculture and Development: A Brief Review of the Literature. Economic Systems. 36. 10.1016/j.ecosys.2011.09.003.

Ditsch,D.C.,R.L.Trimble , and J.M. Wade.1998.Tillage selection:soil stewardship versus financial survival.Specialbulletin,Mississippi Agricultural and Forestry Experiment station 88(1):5 8.

Eastwood, Robert, Michael Lipton and Andrew Newell. 2010. —Farm Size.|| *In Handbook of Agricultural Economics,* Vol. 4, ed. Robert Evenson and Prabhu Pingali. Amsterdam.

Fischer,J., R. Dyball,I. Fazey, C. Gross, S. Dovers, P.R. Ehrlic R.J. Brulle, C. Christensen and R.J. Borden. 2012. Human behavior and sustainability.Frontiers in Ecology and the Environment 10(3);153 160. PB: Washington,USA:Ecological Society oAmerica.URL:http://www.esajournals.org/doi/full/10.1890/110079

Food and Agriculture Organization (FAO), (1991) Sustainable Agriculture and Rural Development in Asia and Pacific, Regional Document No. 2, FAO/ Netherlands Conference on Agriculture and Environment, April 15-19

Foster, Andrew and Mark Rosenzweig (2010). —Barriers to Farm Profitability in India:

GovardhanaRao , G. (2012) An Economic Analysis of Price Behaviour of Rice in Guntur District of Andhra Pradesh.https: / / krishikosh. Egranth.ac.in/

Guha, Sumit(edO(1992) Growth, Stagnation or Decline?-Agricultural Productivity in British india;oxfordUP,Delhi.

Hanamanthrao C. H. (1965) Agricultural Production Functions,Costs and Returns in India;AsiaPublishing House, Bombay

Hazell, P. (2018). Urbanization, agriculture and smallholder farming Agriculture and food systems to 2050: Global trend, challenges and opportunities (pp. 137–160). Singapore:

Headey, Derek and Shenggen Fan. 2008. —Anatomy of a Crisis: The Causes and Consequences of Surging Food Prices.|| Agricultural Economics 39 (1): 375 - 391.

Bacha, S. A. H., Mehwish, S. S., Iqbal, J., Ahmed, A., & Shah, S. (2017). A review on the production and yield of cauliflower in relation with row spacing and various nitrogen levels. *International Journal of Advanced Research and Review,* 2(8), 7-12.

Hussain, Anwar Hussain and khattak, Naeem Ur Rehman Khattak and Khan, Abdul Qayyaum Khan (2008). *Costs Benefit Analysis of Different Rice Varities in District swat*,https: / / krishikosh.egrasnth.ac.in /

ICAR, (1981), Guidelines for entire farming system demonstrations. All India CoordinatedProject on National Demonstrations, Indian Council of AgriculturalResearch, New Delhi: pp.1-12.

Ito, T., & Kurosaki, T. (2009) Weather risk, wages in kind, and the off farmlabour supply of agricultural households in a developing country. American Journal of Agricultural Economics, 91(3), 697–710. https://doi. org/10.1111/j.1467-8276.2009.01270.x

Jha, B. (2004) Implications of Trade Liberalisation for the Livestock Sector, IEG Working PaperSeries No. E/ 243/ 2004

Johnston, Bruce F. and John W. Mellor. 1961. ―The Role of Agriculture in EconomicDevelopment.‖ The American Economic Review 51 (4): 566 – 593.

Kay, p., R. Grayson, M. Phillips , K. Stanley, A. Dodsworth, et al. 2012. The effectiveness of agricultural stewardship for improving water quality at the catchment scale: experiences from an NVZ and ECSFDI watershed.Journal of Hydrology Amsterdam 422 / 423: 10 16.Oxford, UK: Elsevier Ltd.

Kay,P.,A.C.Edwards, and M. Foulger.2009. A review of the efficacy of contemporary agricultural stewardship measures for ameliorating water pollution problems of key concern to the UK water industry.Agricultural systems 99(2/3):67 75.Oxford,UK: Elsevier.

Kingdon, G. G. (2007). The progress of school education in India. Oxford Review of Economic Policy, 23(2), 168–195. https://doi.org/10.1093/oxrep/grm015

Kumar *et al.* (2005).Technical Efficiency of Rice Farms under Irrigated Conditions of North West Himalayan (NWH) Region-A Non-Parametric Approach.https: / / krishikosh.egranth.ac.in /

Kumar, P., R. Gupta &V. C. Mathur (2005). "Agriculture in Future: Dem and-Supply Perspective for theNinth Five Year Plan" Economic and Political Weekly, Vol. 31, No. 39 (September 28).

Kuznets Siman(1970) Economic Growth and Contribution of Agriculture: Note onMeasurement; In: Agriculture inEconomic Development; Edited by Eicher Carl and Witt Lawrexe,Bombay Vora and Company, PageNo. 102 to 119

Kuznets, Simon. 1966. Modern Economic Growth: Rate, Structure, and Spread. New Haven: Yale University Press.

Lanjouw, Jean O. and Peter Lanjouw. (2001). The Rural Non-Farm Sector: Issues and Evidence from Developing Countries.‖ *Agricultural Economics* 26 (1): 1 – 23.

Lanjouw, Jean O. and Peter Lanjouw. 1995. Rural Nonfarm Employment.‖ Policy Research Working Paper No. 1463, World Bank, Washington D.C.

Lewis, W. Arthur. 1954. ―Economic Development with Unlimited Supply of Labour.‖ Manchester School of Economic and Social Studies 22 (2): 139 – 191.

Loayza, Norman V. and Claudio Raddatz. 2010. The Composition of Growth Matters for Poverty Alleviation.‖ Journal of Development Economics 93 (1): 137 – 151.

Markovic, V., M. Djurovka and Z. Ilin. 1990. Influence of different quantity of nitrogen on theyield quality and basic nutritive elements accumulations. Intensive vegetable breedingand greenhouse production (Yugoslavia). p. 81-89

Mechanization, Scale and Credit Markets.|| Paper presented at the Conference *Agriculture for Development-Revisited*, University of California at Berkeley. October 1-2 FAO. (2017). The future of food and agriculture: Trends and Challenges. Rome.

Mehrotra, S. (2014). India's skills challenge: Reforming vocational education and training to harness the demographic dividend. Oxford University Press

Miller ,F.P. (2003). Natural resources management and the second Copernican revolution.Journal of natural resources and Life Sciences Education 32:43: 51,Madison,USA:American Society of Agronomy.

Mishra SS, Acharjee S K, Chakraborty S K, 2009 , Development of tools for assessing conservation categories of siluroid fishes of fresh water and brackish water wetlands of South West Bengal, India.Environ Biol Fish (2009) 84: 395 407@ Springer.

Mundlak, Yair and Donald F. Larson. 1992. —On the Transmission of World Agricultural Prices.|| World Bank Economic Review 6 (3): 399 – 422.

Muralidharan, P. K., (1987), Resource use efficiency in kole lands in Trichur District,Kerala. Indian J. Agric. Econ.,42(4): 578-586.

Patil, B. V., (1992), Sensitivity of field collected Helicoverpaarmigera. Pestology, 17: 19-21

Pingali, P., & Sunder, N. (2017). Transitioning toward nutrition-sensitive food systems in developing countries, Annual Review of Resource Economics, 9(1), 439–459. https://doi.org/10.1146/annurev-resource-100516-053552

Priyanka Chowdhury ,Muvva (2018). Economic analysis of Production Costs and Prices of Rice in west Godavari District of Andhra Pradesh. http: / /krishikosh. egranth.ac.in/

Ray, A. K. and Chahal, J. K., (1986), Problems of oilseeds production. A critical analysis.Agric. Situ. India, 41(5): 323-331

Remenyi, J.N. and Coxhead, I., (1985), Farming systems research in Australia: Results ofa survey. Proceedings on an International Workshop held at Howkesbury, Agric.College, Richmond, N.S. W., (Australia):pp.96-106.

Rome,Italy:Food and agriculture Organization of the United Nations (FAO).2012. FAO Workshop on International Guidelines for Securing Sustainable Small ScaleFisheries,Rome,Italy,7 10 Februrary 2012.FAO Fisheries and Aquaculture Report.2012;(1004): i vii, 1 44

Rosario, E.C., (1985), Farming systems research in Philippines, Proceedings of anInternational Workshop held at Howkesbury, Agric. College, Richmond, N.S. W.,(Australia): pp.119-124

Sabrani, M., Kasryno, F. and Peacock, C., (1985), Farming systems research in Indonesia.Proceedings of an International Workshop held at Howkesbury Agric. College,Richmond N.S. W., (Australia): 116-118.

Sain, I., (1991), A comparative study of wheat paddy versus wheat-cotton farming systemsin the south-western Punjab. Indian J. Agric. Econ., 46(3): 461-462.

Sateesh, A.V. and Sharma, B.M., (1988), Farming systems: Income potential and creditrequirement on non-viable farmers in Pithapuram block. Indian J. Econ. Dev.,13(1): 21-37.

Schultz (1953) , also pointed out the importance of food supply by the agricultural sector. Agriculture is important for economic growth in the sense that it guarantees subsistence for society without which growth is not possible in the first place.

Schultz, Theodore William. 1964. Transforming Traditional Agriculture. New Haven: Yale University Press.Sci., Dharwad., p.10.

Shah, Tushar and Barbara van Koppen, (2006) Is India ripe for integrated water resource management? Fitting water policy to national development context, Economic and Political Weekly, XLI (31) August 5.

Sharma, L.R., Bhati, J.P. and Singh Ranveer, (1991), Emerging farming systems inHimachal Pradesh: Key issues in sustainability. Indian J. Agric. Econ., 46(3): 422-427.

Sharma, V. P. and Singh, R. V., (1996), Economic evaluation of hill cattle developmentprogramme in Himachal Pradesh. Agric. Econ. Res. Rev., 9 (1): 1-12

Shukla, P. (1996). Cost of cultivation and returns on different cost concepts basis of onion in Rajasthan. Economic affairs. 61(1): 11-16

Singh, A.K. 2013. Income and livelihood issues of farmers. Agricultural EconomicsResearch Review, 26: pp 89-96.

Singh, Bibhu Ranjan Kr. (2017). Study on *Cultivation Practices and marketing of rice in Palamau District of Jharkhand.* https:/ /krishikosh.egranth.ac.in/

Srinivasan, P.V. and Shikha Jha. 2001. ―Liberalized Trade and Domestic Price Stability. The Case of Rice and Wheat in India.‖ *Journal of Development Economics* 65 (2): 417 – 441.

Stonehouse,D.P.(2004).Sustainabilty issues in the agri food sector in Ontario,Canada. *Journal of Sustainable Agriculture* 23(3):109 124,Binghamton, USA:Food Products Press.

Sundaramurthy, V. T., (1985), Operational research project on integrated control of pests.Annual Report, Central Institute of Cotton Research, Nagpur, pp. 111-112.

Sundaramurthy, V. T., (1987), Operational research project on integrated control of pests. Annual Report, Central Institute of Cotton Research, Nagpur, pp. 56-77.

Suresh, T.R. Keshava Reddy (2006). *Resource-use efficiency of Paddy cultivation in Peechi Command Area of Thrissur District of Kerala: An Economic Analysis.* https: / / krishikosh.egranth.ac.in/

Suryawanshi, S. D. and Mahindre Prakash, (1993), Impact of viable technology for promoting oilseeds in Maharashtra. Indian J. Agric. Econ., 48(3): 420

Timmer, C. Peter. 2002. —Agriculture and Economic Development.|| In Handbook ofAgricultural Economics, Vol. 2A, ed. Bruce L. Gardner and Gordon C. Rausser .Amsterdam: North Holland (10) (PDF) *Agriculture and Development: A Brief Review of the Literature.*

Vakati, Narendra (2015). Economics of production and marketing of paddy in Nellore district of Andhra Pradesh. *https: / / krishikosh.egranth.ac.in/*

Vishweshwar, S.P., (1994), Economics of hybrid cotton with special reference to pestmanagement in Malaprabha command area. M.Sc. (Agri.) Thesis, Univ. Agric.

Von Braun, Joachim and Maximo Torero. 2009. —Implementing Physical and Virtual Food Reserves to Protect the Poor and Prevent Market Failure.|| Policy Brief 10, IFPRI, Washington D.C.

Vyas, V.S. and V. Ratna Reddy, (1996) Environmental policies and policy implementation in India: An assessment, In: Environment and Development, Ed: M.S. Rathore. Jaipur and New Delhi: Rawat Publication.

Webb, Patrick and Thomas Reardon. 1992. —Drought Impact and Household Response in East and West Africa.|| *Quarterly Journal of International Agriculture* 31 (3): 230 – 246.

Wodon, Quentin and Hassan Zaman. 2008. Rising Food Prices in Sub-Saharan Africa: Poverty Impact and Policy Responses.|| Policy Research Working Paper No. 4738, World Bank, Washington D.C.

World Bank. 2003. Land Policies for Growth and Poverty Reduction. Washington D.C. and Oxford: World Bank and Oxford University Press

World Bank. 2008. —Rising Food Prices: Policy Options and World Bank Response.|| Washington D.C.

Wright, Brian. 2009. —International Grain Reserves and Other Instruments to Address Volatility in Grain Markets.|| Policy Research Working paper No. 5028, World Bank, Washington D.C.

INDEX

A

Agribusiness 12
Agricultural processes 5
Agriculture 1
Agriculture hubs 12
Agriculture trends 22
Alluvial plain 41
Alluvial zone 80
Aman rice 43
Analytic tools 66
Anthocyanin pigments 7
Arable land 40
Asia's dominance 8

B

Biodiversity 22
Biodiversity conservation 134
Bio energy crops 20

C

Calorie content 7
Civil society organizations 34
Climate change affects 22
Cluster analysis 73, 138
Coastal harvest 9
Crop cultivation 44
Crop ecology 37
Cropping intensity 63
Cropping pattern 43
Crop production 10
Cyclone 40

D

Demography 37
Deprivation 15
Development strategy 11

E

Ecological services 143
Economic opportunities 11
Eigen value 71
Eigenvector 72
Emerging modern food 12
Employment growth 12
Energy management 109
Environmental problems 17
Euclidean distance 119
Exogenous variables 136

F

FaaS 12
Factor analysis 69
Family education score 62
Farm economy 13
Farm income 90, 135
Farming system 5, 6
Farm laboring 10
Fertilizer-use 21

Field crop 104
Fish entrepreneur 143
Fishing industry 9
Fish trading 10
Food and agriculture organization (FAO) 17
Food security 10, 11
Food supply chains, 24
Formal water economy 19
Fragmentation 109

G

GDP declines 13
Global economy 6
Global food 6
Global market 129
Gold farm 12
Greenhouse gases 19
Green revolution 11

H

Hailstorm 40
Haringhata block 46
Haringhata dairy farm 49
Healthcare 50
Homestead land 63, 86, 118
Homogenous 120
Human development report 48
Humans civilization 118

I

Indian agriculture 1, 101, 129
Individual farm systems 6
Industry 44
Infrastructure 48
Integrated Water Resources Management (IWRM) 18
Isochronously dealt 128
Income diversification 11

J

Job creation 11
Jobs agenda 11
Job skills 13
Jute industry 45

K

Kharif 9

L

Land degradation 22
Land tenure 33
Linking food systems 11
Livestock 5
Long-grain rice 7

M

Marketing system 15
Mean 66
Mechanization 12
Methaneemission 134
Methane gas 21

N

Natural resource 22, 108
Nutrition 11

P

Paddy fields 21
Petroleum products 19
Pilot study 61
Pioneer crop 8
Pollution 39
Principal components. 72
Prototype agriculture 129
Public sector's role 20
Purple rices 7
Principal food crop 26

R

Rabi 9
Research methodology 59
Rice production 7
Rice yields 8

S

Skill enhancement 12
Smallholder sector 33
Social-ecological 3
Social security 13
Soil organic carbon 60
Sorrow of bengal 39
Standard deviation 66
Sub-saharan africa (SSA) 15
Sustainable agriculture 15
Sustainable development goals (SDGS) 14

T

Technological changes 17
Theoretical orientation 5
Trade-offs 19
Transport 45

U

Unimodal 11

W

Water fisheries 2
Water poverty 18
Water resources 18
World bank 12
World's food system 17

Printed at Repro India Ltd.